Math Mammoth
Grade 4 Answer Keys

for the complete curriculum
(Light Blue Series)

Includes answer keys to:

- Worktext part A
- Worktext part B
- Tests
- Cumulative Reviews

By Maria Miller

Contents

Answer Key, Worktext part A 5

Answer Key, Worktext part B 57

Answer key, Chapter 1 Test 107
Answer key, Chapter 2 Test 107
Answer key, Chapter 3 Test 107
Answer key, Chapter 4 Test 108
Answer key, Chapter 5 Test 108
Answer key, Chapter 6 Test 108
Answer key, Chapter 7 Test 109
Answer key, Chapter 8 Test 110
Answer key, End-of-the-Year Test 110

Answer key, Cumulative Review Chapters 1-2 117
Answer key, Cumulative Review Chapters 1-3 117
Answer key, Cumulative Review Chapters 1-4 118
Answer key, Cumulative Review Chapters 1-5 119
Answer key, Cumulative Review Chapters 1-6 120
Answer key, Cumulative Review Chapters 1-7 121
Answer key, Cumulative Review Chapters 1-8 123

Math Mammoth Grade 4-A
Answer Key

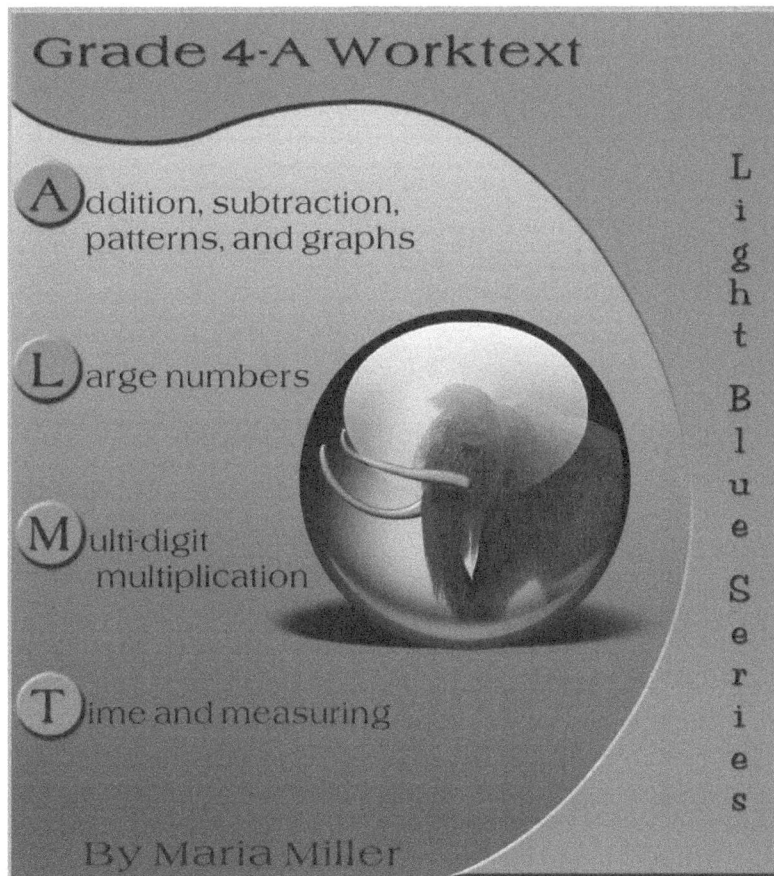

By Maria Miller

Math Mammoth Grade 4-A Answer Key
Contents

	Work-text page	Answer key page

Chapter 1: Addition, Subtraction, Patterns, and Graphs

Addition Review	11	8
Adding in Columns	14	8
Subtraction Review	15	8
Subtract in Columns	18	9
Patterns and Mental Math	21	9
Patterns in Pascal's Triangle	23	10
Bar Models in Addition and Subtraction	25	10
Order of Operations	29	11
Making Bar Graphs	31	12
Line Graphs	33	12
Rounding	36	14
Estimating	39	14
Money and Discounts	41	15
Calculate and Estimate Money Amounts	44	16
Review Chapter 1	47	16

Chapter 2: Large Numbers and Place Value

Thousands	51	17
At the Edge of Whole Thousands	54	17
More Thousands	56	18
Practicing with Thousands	58	18
Place Value with Thousands	60	19
Comparing with Thousands	62	19
Adding ands Subtracting Big Numbers	65	20
Rounding and Estimating with Large Numbers	69	21
Multiples of 10, 100, and 1000	73	23
Mixed Review Chapter 2	76	24
Review Chapter 2	78	25

Chapter 3: Multiplication

Understanding Multiplication	84	26
Multiplication Tables Review	87	26
Scales Problems	90	28
Multiplying by Whole Tens and Hundreds	94	28

Chapter 3: Multiplication (cont.)

Multiply in Parts 1	98	29
Multiply in Parts 2	101	30
Multiply in Parts—Area Model	103	31
Multiplying Money Amounts	105	32
Estimating in Multiplication	107	33
Multiply in Columns—the Easy Way	109	33
Multiply in Columns—the Easy Way, Part 2	112	34
Multiply in Columns, the Standard Way	115	34
Multiplying in Columns, Practice	119	35
Order of Operations Again	121	35
Money and Change	124	36
So Many of the Same Thing	127	37
Multiplying Two-Digit Numbers in Parts	130	38
Multiplying by Whole Tens in Columns	135	40
Multiplying in Parts: Another Way	137	41
The Standard Multiplication Algorithm with a Two-Digit Multiplier	139	41
Mixed Review Chapter 3	143	42
Review Chapter 3	145	42

Chapter 4: Time and Measuring

Time Units	153	44
Elapsed Time 1	156	44
The 24-Hour Clock	159	45
Elapsed Time 2	161	45
Elapsed Time 3	164	46
Measuring Temperature: Celsius	167	46
Measuring Temperature: Fahrenheit	171	48
Temperature Line Graphs	173	48
Measuring Length	175	49
More Measuring in Inches and Centimeters	178	49
Feet, Yards, and Miles	180	49
Metric Units for Measuring Length	185	50
Customary Units of Weight	188	51
Metric Units of Weight	192	51
Customary Units of Volume	195	52
Metric Units of Volume	198	53
Mixed Review Chapter 4	201	54
Review Chapter 4	203	54

Chapter 1: Addition, Subtraction, Patterns, and Graphs

Addition Review, p. 11

1. a. 150, 157, 159 b. 190, 191, 199
   c. 110, 119, 120 d. 170, 175, 179

2. a. 400 + 80 + 7 b. 2,000 + 100 + 3
   c. 8,000 + 40 + 5 d. 600 + 50

3. a. It was 44. 56 + 90 + 44 = 190
   b. 70 + 80 = 150

4. a. 15, 65, 150, 1500
   b. 13, 43, 130, 330
   c. 14, 24, 1400, 640

5. For example 50 + 80 = 130;
   500 + 800 = 1,300; 25 + 8 = 33

6. a. 87 + 34 + 44 = 165, 5 + 2 + 4 = 11,
   154 + 11 = 165
   b. 127 + 500 + 90 = 717, 4 + 3 + 9 = 16,
   717 + 16 = 733

7. Add one hundred then subtract one.
   a. 56 + 100 = 156; 156 − 1 = 155,
   b. 487 + 100 = 587; 587 − 1 = 586

8. a. 153, 79, 121 b. 89, 128, 111 c. 181, 101, 149

9.

Half the number	_10_	45	55	60	240	450	800	2,005
Number	20	90	110	120	480	900	1,600	4,010
Its double	_40_	180	220	240	960	1,800	3,200	8,020

10. a. $116 b. $105

11.

n	56	156	287	569	950	999
$n + 999$	1,055	1,155	1,286	1,568	1,949	1,998

12. a. 1,200, 1,800, 2,400, 3,000, 3,600, 4,200
    It reminds me of the multiplication table of 6.

    b. 1,800, 2,700, 3,600, 4,500, 5,400, 6,300
    It reminds me of the multiplication table of 9.

    c. 175, 250, 325, 400, 475, 550, 625, 700

Adding in Columns, p. 14

1. a. 5,539 b. 9,058 c. 8,683

2. a. 8,325 b. 5,657

3. a. 672 miles b. 261 miles

Subtraction Review, p. 15

1. a. 6, 56 b. 6, 76 c. 6, 60 d. 8, 800

2. a. 98, 80, 78 b. 196, 160, 155
   c. 495, 450, 444 d. 393, 330, 329

3. a. 4, 34, 40, 440
   b. 6, 66, 60, 560
   c. 6, 66, 600, 360

4. Answers will vary. For example: 34 − 8 = 26,
   140 − 80 = 60, 240 − 80 = 160,
   and 740 − 80 = 660.

5.

n	125	293	404	487	640	849
$n - 99$	26	194	305	388	541	750

6. a. 9, 5, 13 b. 18, 44, 48
   c. 27, 22, 46 d. 70, 50, 440
   e. 445, 944, 792 f. 418, 542, 492

7. a.

n	120	140	160	180	200	220	240	260	280
$n - 27$	93	113	133	153	173	193	213	233	253

b. Each answer ends in 3, and each answer is 20 more
   than the previous answer.

8. a. 240, 200, 160, 120, 80, 40.
   It reminds me of the multiplication table of 4.
   b. 5,400, 4,800, 4,200, 3,600, 3,000, 2,400.
   It reminds me of the multiplication table of 6.
   c. 490, 420, 350, 280, 210, 140.
   It reminds me of the multiplication table of 7.

9. Game:
   a. 21 − 5 − 5 − 5 − 5 = 1
   b. 37 − 10 − 10 − 10 = 7
   c. 37 − 12 − 12 − 12 = 1 and
   50 − 7 − 7 − 7 − 7 − 7 − 7 − 7 = 1
   d. 30 − 9 − 9 − 9 = 3 and 20 − 8 − 8 = 4.

Subtract in Columns, p. 18

1. a. 173   b. 3,809   c. 568
   d. 344   e. 3,764   f. 5,326
   g. 217   h. 305   i. 5,580

2. a. 1,162   b. 4,925

3. First add 592, 87, 345 and 99; then subtract the sum from 5,200. The answer is 4,077

4. a. 1,530 miles. One round trip is 255 + 255 = 510 miles. Three round trips are 510 + 510 + 510 = 1,530 miles.

   b. 74 miles longer.

Puzzle Corner: 4:25

Patterns and Mental Math, p. 21

1. a.

n	9	18	27	36	45	54	63	72	81	90
$n+29$	38	47	56	65	74	83	92	101	110	119

   b. The skip-counting pattern by 9's.
   c. Yes. There is a skip-counting pattern by 9's in the bottom row too, but it starts at 38.

2. a. *Hint: instead of subtracting 39, subtract 40, and add 1!*

n	660	600	540	480	420	360	300	240
$n-39$	621	561	501	441	381	321	261	201

   b. It is a skip-counting pattern by 60's, backwards.
   c. Yes. It also has a skip-counting pattern going backwards by 60's.

3. a. 497, 470, 200, 467, 197
   b. 598, 580, 400, 578, 398
   c. 993, 930, 300, 923, 293

4.

5. Subtract a thousand, then add one. To do 1,446 − 999, first subtract a thousand: 1,446 − 1,000 = 446. Then add one: 446 + 1 = 447.

6. a. The second alarm clock cost $11 + $8 = $19, and $11 + $19 = $30.
   b. June has 30 days and July has 31. There are 25 days with no rain in June and 25 days in July; a total of 50 days.
   c. The difference is 162 cm − 134 cm = 28 cm.
   d. Jack rode 100 km in all. 28 + 28 + (28 − 6) + (28 − 6) = 100 km
   e. There are 45 − 18 = 27 boys, and 27 − 18 = 9 more boys than girls.

Patterns in Pascal's Triangle, p. 23

1.

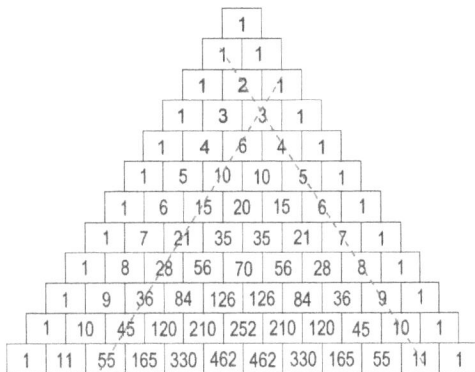

2. The row sums are:
   1, 2, 4, 8, 16, 32, 64, 128, 256, 512, 1024, 2048.
   These numbers double each time.

3. Yes - it is marked with a dashed line in the image.

4. a. 1, 3, 6, 10, 15, 21, 28, 36, 45, 55
   b. The diagonal is marked with a dashed line in the image.
   c. When we find the differences of neighboring numbers, we get:
   2, 3, 4, 5, 6, 7, 8, 9, 10, 11, and so on—which are the counting numbers.

Bar Models in Addition and Subtraction, p. 25

1.

a. $128 + x = 400$
    $x = 400 - 128 = 272$
	b. $x + 385 = 999$
	    $x = 999 - 385 = 614$

c. $\$890 + x = \$1,200$
   $x = \$1,200 - \$890 = \$310$

$\longleftarrow \$1,200 \longrightarrow$

$\$890$	x

d. Let x be the number of boys.

   $265 + x = 547$
   $x = 547 - 265 = 282$

$\longleftarrow$ total 547 $\longrightarrow$

265	x

2.

a.

$\vdash\!\!\!-\!\!\!- 1,200 -\!\!\!-\!\!\!\dashv$

420	370	x

Addition: $420 + 370 + x = 1,200$
Solution: $x = 410$

b.

$\vdash\!\!\!-\!\!\!- 4,000 -\!\!\!-\!\!\!\dashv$

400	400	400	x

Addition: $400 + 400 + 400 + x = 4,000$
Solution: $x = 2,800$

c.

$\vdash\!\!\!-\!\!\!- 250 -\!\!\!-\!\!\!\dashv$

8	x	8

Addition: $28 + 28 + x = 250$
Solution: $x = 194$

d.

$\vdash\!\!\!-\!\!\!- x -\!\!\!-\!\!\!\dashv$

56	9	118

Addition: $56 + 9 + 118 = x$
Solution: $x = 183$ miles

3. Answers will vary. For example: A swimming pool costs $4,900. A family has saved $1,750 for it.
   How much more do they still need to save? $x + 1,750 = 4,900$, from which $x = 4,900 - 1,750 = 3,150$.

4. a. $x - 29 = 46$; $x = 29 + 46 = 75$
   b. Answers will vary. For example: $x - 255 = 99$ OR $x - 99 = 255$; $x = 255 + 99 = 354$.

5. a. 24   b. 32   c. 99   d. 15   e. 72   f. 475

6.

a. $\xleftarrow{\qquad}$ 52 $\xrightarrow{\qquad}$ 24 | 28 $52 - x = 28$ $x = 52 - 28 = 24$	b. $\xleftarrow{\qquad}$ 97 $\xrightarrow{\qquad}$ 43 | 54 $97 - x = 54$ $x = 97 - 54 = 43$

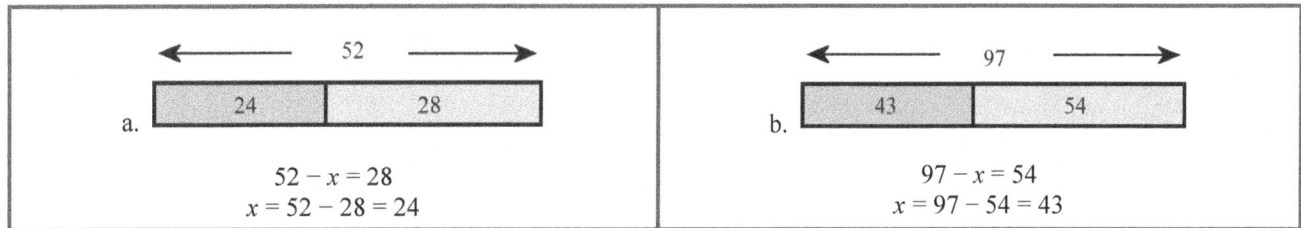

7. a. 8   b. 21   c. 134   d. 18   e. 28   f. 557

8. a. $\$15 + x = \$22$; $x = \$7$
   b. $x - 24 = 125$; $x = 149x$
   c. $120 - \underline{\quad} = 89$; $x = 31$
   d. $x - 67 = 150$; $x = 217$

9. a. $x + 43 = 450$. Subtract to solve x. $x = 450 - 43 = 407$
   b. $\$250 - x = \$78$. Subtract to solve x. $x = 250 - 78 = \$172$
   c. $\$200 - \$54 - \$78 = x$. Subtract to solve x. $x = \$68$
   d. $x - \$23 - \$29 = \$125$. Add to solve x. $x = \$125 + \$29 + \$23 = \$177$

Puzzle Corner: a. 60   b. 220   c. $x = 31$ d. $x = 37$

In (a), we know the total (subtraction always starts with the total), and one of the PARTS is missing. To find the missing part, subtract the other parts from the total. So, to solve $200 - 45 - \underline{\quad} - 70 = 25$, subtract the other parts (45, 70, and 25) from 200.

In (b), the total is missing: $\underline{\quad} - 5 - 55 - 120 = 40$. We can find it by adding all the parts (the 5, 55, 120, and 40).

The problems in (c) and (d) have a missing addend. They are solved by subtracting all the parts from the total.

Order of Operations, p. 29

1. a. 16   b. 13   c. 19
   d. 11   e. 22   f. 14
   g. 4   h. 5   i. 49

2. $90 - 2 \times 20 = 50$. A 50-cm piece is left.

3. $100 \text{ kg} - 4 \times 5 \text{ kg} = 80 \text{ kg}$
   or $100 \text{ kg} - 5 \text{ kg} - 5 \text{ kg} - 5 \text{ kg} - 5 \text{ kg} = 80 \text{ kg}$
   or $100 \text{ kg} - (5 \text{ kg} + 5 \text{ kg} + 5 \text{ kg} + 5 \text{ kg}) = 80 \text{ kg}$

4. $5 \times \$2 + 2 \times \$3 = \$16$ *or* $2 \times \$3 + 5 \times \$2 = \$16$
   or $\$2 + \$2 + \$2 + \$2 + \$2 + \$3 + \$3 = \$16$

5. a. $4 \times 1 + 8 = 12$ or $4 + 1 \times 8 = 12$
   b. $2 + 10 + 1 \times 2 = 14$ OR $2 + 10 \times 1 + 2 = 14$
   c. $3 \times 3 - 3 = 6$

6. a. The first and the last problem have the same answer.
   b. The first and the second problem have the same answer.
   c. The first and the last problem have the same answer.

7. a. and c.

8. Answers will vary. For example:
   Tim bought four ice cream cones for $1.20 each, and paid with $10. What was his change?
   $\$10 - 4 \times \$1.20 = \$5.20$

9. a. 26; 191   b. 11; 100   c. 33; 14

10. a. $50 - 5 \times 10 = 0$ or $50 \div 5 - 10 = 0$;
    b. $100 - (15 + 17) \times 1 = 68$
    c. $(2 + 5) \times 2 = 14$

Making Bar Graphs, p. 31

1. a.

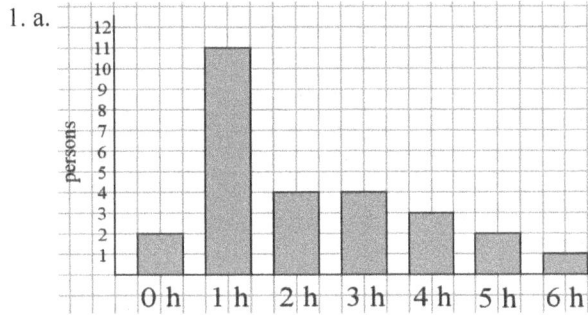

Hours of TV	Frequency
0 h	2
1 h	11
2 h	4
3 h	4
4 h	3
5 h	2
6 h	1

b. 27 classmates   c. 1 hour of TV.
d. 13   e. 10
f. no   g. yes.

2. a.

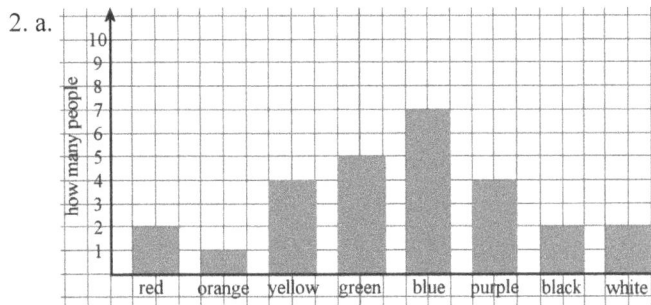

b. 27 people   c. "Cold" colors.

3. a.

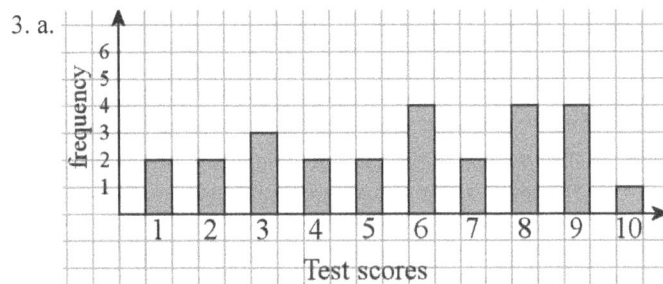

Test score	Frequency
1	2
2	2
3	3
4	2
5	2
6	4
7	2
8	4
9	4
10	1

3. b. 6, 8, and 9; four students each
   c. 10; 1 student
   d. 12
   e. 5
   f. 9

Line Graphs, p. 33

1. a. $60
   b. $140
   c. June
   d. $15
   e. She had $70 less than in August. Perhaps she
      purchased something with her savings.

2. a. Day 1: 500 grams; Day 2: 525 grams;
      Day 3: 550 grams; Day 4: 575 grams
   b. Day 5.
   c. Day 8.

3. a. The price lowers in the summer and is higher in
      the winter. That is because in the summer there is an
      abundance of strawberries; all stores and markets are
      selling strawberries. Nobody can keep the price high
      because if they did, people would go elsewhere to buy.
   b. The highest price was in December, $3.60 per pound,
      and the lowest price was in July, $1.63 per pound.
      The difference is $1.97.
   c. In August, $3.64. In November, $6.38.

4. a.

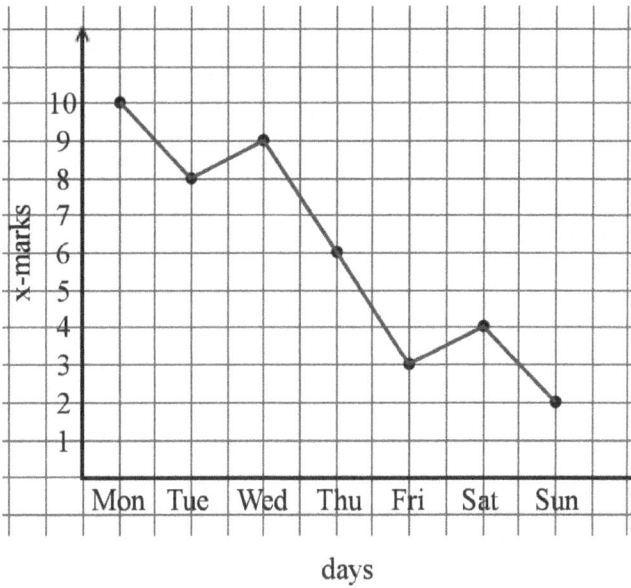

days

b. yes.

5. a.

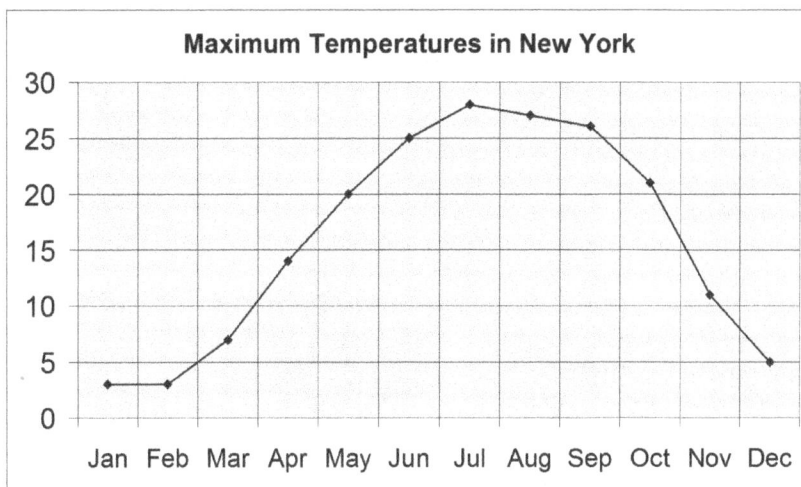

Maximum Temperatures in New York

b. January, February, and December
c. June, July, August, and September
d. 25 degrees

Rounding, p. 36

1. a. 290 b. 320 c. 280 d. 290 e. 320 f. 300 g. 300 h. 210

2. a. 530 b. 30 c. 180 d. 200 e. 710 f. 390
   g. 440 h. 5,970 i. 9,570 j. 4,060 k. 2,280 l. 4,000

3. a. 3,500 b. 3,700 c. 3,900 d. 3,500 e. 4,000 f. 3,400

4. a. 500 b. 100 c. 800 d. 200 e. 700 f. 400
   g. 2,900 h. 6,000 i. 7,500 j. 3,000 k. 3,000 l. 4,000

5. a. 4,000 b. 7,000 c. 5,000 d. 7,000 e. 3,000 f. 4,000

6. a. 1,000 b. 0 c. 1,000 d. 4,000 e. 6,000 f. 3,000
   g. 3,000 h. 6,000 i. 9,000 j. 10,000 k. 3,000 l. 1,000

7.

n	55	2,602	9,829	495	709	5,328
rounded to nearest 10	60	2,600	9,830	500	710	5,330
rounded to nearest 100	100	2,600	9,800	500	700	5,300
rounded to nearest 1000	0	3,000	10,000	0	1,000	5,000

Estimating, p. 39

1. a. Estimation: $1,000 + 200 + 4,800 = 6,000$ Exact: 5,990
   b. Estimation: $300 + 400 + 600 = 1,300$ Exact: 1,312
   c. Estimation: $1,000 - 400 - 100 = 500$ Exact: 552
   d. Estimation: $3,500 - 1,500 - 200 = 1,800$ Exact: 1,741

2. About $4 \times 50 = 200$ passengers.

3. About $150 + $160 + $180 + $130 + $130 = $750

4. a. Apartment 1: about $290 + $290 + $290 = $870
      Apartment 2: about $330 + $330 + $330 = $990.
   b. They would save approximately: $990 − $870 = $120.

5. a. $340 + 360 + 320 + 320 = 1,340$
   b. $300 + 290 + 290 + 260 = 1,140$

1. a. 25¢   b. 178¢   c. 1560¢
   d. $0.20   e. $1.54   f. $8.59

2. a. $2.20 + $12 + $1.50 = $15.70
   b. $20 − $2.20 − $12 − $1.50 = $4.30

3.

Item cost	Money given	Change needed	$50 bill	$20 bill	$5 bill	$1 bill
a. $56	$70	$14			2	4
b. $78	$100	$22		1		2
c. $129	$200	$71	1	1		1

4.

Item cost	Money given	Change needed	$5 bill	$1 bill	25¢	10¢	5¢	1¢
a. $2.56	$5	$2.44		2	1	1	1	4
b. $7.08	$10	$2.92		2	3	1	1	2
c. $3.37	$10	$6.63	1	1	2	1		3

5. a. **ii.** $x = \$12$ b. **iii.** $x = \$108$ c. **i.** $12

6. b. $x − \$250 = \$170$; $x = \$420$

   c. $x − \$45 = \$15$; $x = \$60$

   d. $\$12 − x = \$3.56$; $x = \$8.44$

   e. $x − \$12 − \$9 = \$29$; $x = \$50$

   f. $\$65 − \$12 − \$12 − \$7 = x$; OR $\$65 − 2 \times \$12 − \$7 = x$; $x = \$34$

   g. $\$20 − x − x = \$12.40$ OR $\$20 − 2x = \$12.40$; $x = \$3.80$

   h. $\$50 − x − x − x = \$17$ OR $\$50 − 3x = \$17$; $x = \$11$.

7. $999

8. $19

9. a. $0.75
   b. $0.24
   c. $477/month
   d. $35

10. a. $\$54.99 − x = \$47.99$; $x = \$7$

    b. $x − \$35 = \$94$; $x = \$129$

    ←—— original price $129 ——→

$94	$35

Calculate and Estimate Money Amounts, p. 44

1. a. $6.30 b. $10.00 c. $5.60 d. $0.30 e. $0.70 f. $5.00

2. a. $3 b. $98 c. $3 d. $1,680 e. $47 f. $126

3. a. $50 b. $10 c. $70 d. $6,290 e. $40 f. $170

4.

n	$29.78	$5.09	$59.95	$2.33	$0.54
rounded to nearest ten cents	$29.80	$5.10	$60.00	$2.30	$0.50
rounded to nearest dollar	$30.00	$5.00	$60.00	$2.00	$1.00

5. a. $2 + $6 + $5 + $13 = $26
   b. $120 + $200 + $260 + $340 = $920

6. Answers will vary because the exact ways of rounding will vary.
   a. about $12 (rounding to the nearest dollar) or about $11.90 (rounding to the nearest ten cents)
   b. six gallons
   c. about 5 × $2 + 2 × $5 = $20
   d. five ice cream cones

7. a. $23.30 b. $369.50 c. $201.01 d. $30.75

8. a. A 1-Day Adult ticket costs $6 more than a 1-Day Child ticket.
   b. A 2-Day Adult ticket costs $12 more than a 2-Day Child ticket.
   c. For a 4-Day Adult ticket, the discount is $2.48.
   d. For a 4-Day Child ticket, the discount is $3.32.

9. a. The tickets would cost a total of $929.58.
   b. Yes, they can. The total for 4-day discount tickets would be $978.40.

Review Chapter 1, p. 47

1.

a. 81 − 72 = 9 665 − 99 = 566	b. 45 + 65 = 110 196 + 99 = 295	c. 160 + 280 = 440 54 − 28 = 26

2. $x + 38 = 230$; $x = 192$ ← total 230 →

192	38

3. $x + 587 = 1,394$; $x = 1,394 − 587 = 807$

4. a. 30, 70
   b. 100, 29
   c. 82, 76

5. ($13 − $2) × 3 = $33.

6. (10 × 4) + (20 × 2) = 80 feet altogether.

7. $25 + $14 + $3 = $42

8. $15.20 + $34.60 + $70.20 = $120

9. $48.90 + ($48.90 + $25) = $122.80

Chapter 2: Large Numbers and Place Value

Thousands, p. 51

1. b. $4,000 + 900 + 30 + 5$ c. $4,000 + 0 + 30 + 9$ d. $3,000 + 0 + 0 + 2$
   e. $2,000 + 0 + 90 + 0$ f. $9,000 + 400 + 0 + 5$

2. a. 4,593   b. 2,090   c. 3,200   d. 8,005   e. 4,600   f. 4,080
   g. 7,203   h. 1,405   i. 7,050   j. 4,005   k. 4,069   l. 3,809

3. b. five hundred   c. five thousand   d. fifty

4. b. nine thousand   c. forty   d. eighty   e. two hundred   f. two   g. twenty   h. five

5. a. 8,542   b. 2,458

6. The difference is $961 - 169 = 792$.

7.

n	2,508	342	4,009	59	6,980	8,299
$n + 10$	2,518	352	4,019	69	6,990	8,309
$n + 100$	2,608	442	4,109	159	7,080	8,399
$n + 1000$	3,508	1,342	5,009	1,059	7,980	9,299

8. a. 6   b. 400   c. 8   d. 8,000   e. 5,000   f. 600

9. 7,889

Puzzle Corner.   b. 29 and 92; difference: 63
   c. 45 and 54; difference: 9
   d. 38 and 83; difference: 45
   e. You can find all those in the multiplication table of 9.
   f. 0 and 4, 1 and 5, 2 and 6, 3 and 7, 4 and 8, 5 and 9
   g. 0 and 3 $(30 - 3 = 27)$, 1 and 4, 2 and 5, 3 and 6, 4 and 7, 5 and 8, or 6 and 9
   h. 0 and 9

At the Edge of Whole Thousands, p. 54

1. a. $991 + 9 = 1,000$ b. $960 + 40 = 1,000$ c. $942 + 58 = 1,000$
   d. $924 + 76 = 1,000$ e. $933 + 67 = 1,000$ f. $979 + 21 = 1,000$

2. a. $999 + \underline{1} = \underline{1,000}$; $992 + \underline{8} = \underline{1,000}$
   b. $980 + \underline{20} = \underline{1,000}$; $985 + \underline{15} = \underline{1,000}$
   c. $930 + \underline{70} = \underline{1,000}$; $937 + \underline{63} = \underline{1,000}$

3. a. $1,920 + 80 = 2,000$; $1,999 + 1 = 2,000$; $2,998 + 2 = 3,000$
   b. $1,990 + 10 = 2,000$; $7,940 + 60 = 8,000$; $5,970 + 30 = 6,000$
   c. $6,950 + 50 = 7,000$; $4,900 + 100 = 5,000$; $3,995 + 5 = 4,000$

4. a. 1,999; 1,996; 1,993   b. 4,997; 3,990; 6,980   c. 5,950; 8,970; 9,900

5.

Number	Rounded number	Rounding error
4,993	5,000	7
7,890	8,000	110
9,880	10,000	120

Number	Rounded number	Rounding error
8,029	8,000	29
5,113	5,000	113
2,810	3,000	190

6. a. 1,900; 1,850; 1,750
   b. 4,800; 4,770; 4,720
   c. 8,500; 8,420; 8,320

7. a. $3,000   b. It is $3,000 - $2,992 = $8 short.

8. The error is 68. The first addend, 1982, is 18 less than 2,000, and the second addend, 3,950 is 50 less than 4,000.
   This creates a total rounding error of $18 + 50 = 68$.

More Thousands, p. 56

1. a. 164,000; 164 thousand
   b. 92,000;   92 thousand
   c. 309,000; 309 thousand
   d. 34,000;   34 thousand
   e. 780,000; 780 thousand

2. b. 92,908; 92 thousand 908   c. 329,033; 329 thousand 033
   d. 14,004; 14 thousand 004   e. 550,053; 550 thousand 053
   f. 72,001; 72 thousand 001   g. 800,004; 800 thousand 004
   h. 30,036; 30 thousand 036

3. a. four hundred fifty-six thousand ninety-eight
   b. nine hundred fifty thousand fifty
   c. twenty-three thousand ninety
   d. five hundred sixty thousand eight
   e. seventy-eight thousand three hundred four
   f. two hundred sixty-six thousand eight hundred ninety-four
   g. one million
   h. three hundred six thousand seven hundred

4. a. 35,000   b. 201,000   c. 430,000   d. 750,000   e. 1,000,000   f. 770,000

5. a. 40,000   b. 721,000   c. 450,000   d. 630,000   e. 240,000
   f. 800,000   g. 25,000   h. 194,000   i. 323,000   j. 499,000

6. The blue dots on the number line mark the numbers 502,000, 511,000, 524,000, and 538,000.

7.

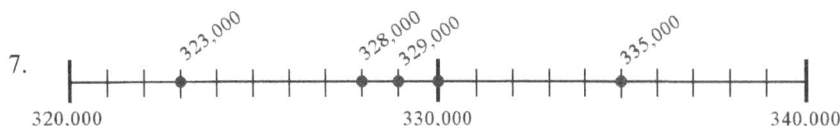

Practicing with Thousands, p. 58

1. a.   49 thousands 0 hundreds 1 ten 5 ones   b. 206 thousands 0 hundreds 9 tens 0 ones
   c. 107 thousands 8 hundreds 0 tens 2 ones   d.   88 thousands 0 hundreds 3 tens 0 ones
   e. 790 thousands 3 hundreds 0 tens 2 ones   f. 903 thousands 0 hundreds 0 tens 0 ones
   g. 250 thousands 0 hundreds 6 tens 7 ones   h. 300 thousands 0 hundreds 7 tens 0 ones

2. a. 20,704   b. 204, 080   c. 101,600   d. 540,004   e. 230,370
   f. 9,607   g. 873,050   h. 40,400   i. 59,065

3. a. 25,347   b. 700,624   c. 61,808   d. 53,060   e. 42,087
   f. 1,000,000   g. 290,040   h. 27,905   i. 504,008

4.

a.	b.	c.	d.
45,000	134,000	800,000	400,000
45,500	134,200	750,000	390,000
46,000	134,400	700,000	380,000
46,500	134,600	650,000	370,000
47,000	134,800	600,000	360,000
47,500	135,000	550,000	350,000
48,000	135,200	500,000	340,000
48,500	135,400	450,000	330,000
49,000	135,600	400,000	320,000

5. a. 30,050   b. 254,305   c. 133,250   d. 77,004   e. 60,002
   f. 120,063   g. 15,020   h. 24,006   i. 30,390   j. 86,471

Place Value with Thousands, p. 60

1. a.

hth	tth	th	h	t	o
	8	7	0	1	5
	8	0	0	0	0
		7	0	0	0
				1	0
					5

b.

hth	tth	th	h	t	o
4	0	3	2	8	0
4	0	0	0	0	0
		3	0	0	0
			2	0	0
				8	0

c.

hth	tth	th	h	t	o
6	9	2	0	0	4
6	0	0	0	0	0
	9	0	0	0	0
		2	0	0	0
					4

d.

hth	tth	th	h	t	o
7	0	0	2	0	4
7	0	0	0	0	0
			2	0	0
					4

Optionally, you can add a row of zeros for the places where the number has a zero, like this:

d.

hth	tth	th	h	t	o
7	0	0	2	0	4
7	0	0	0	0	0
0	0	0	0	0	0
	0	0	0	0	0
			2	0	0
				0	0
					4

2. a. 80,000 + 7,000 + 10 + 5
   b. 400,000 + 3,000 + 200 + 80

3. a. 30,000 + 2,000 + 400 + 90 + 3
   b. 100,000 +70,000 + 2,000 + 300 + 90 + 2
   c. 20,000 + 5,000 + 600
   d. 100,000 + 9,000 + 20
   e. 900,000 + 700 + 1

4. a. 20,000   b. 5,000   c. 20   d. 7,000   e. 70,000

5. b. five   c. five hundred   d. fifty thousand

6. a. ten thousand   b. three hundred thousand   c. eighty thousand   d. eight thousand
   e. six hundred   f. twenty   g. ten thousand   h. nine hundred thousand

7. a. 500,087   b. 20,480   c. 907,700   d. 50,360

Puzzle corner. 630,001

Comparing with Thousands, p. 62

1. a. <   b. <   c. <   d. >   e. <   f. <   g. <   h. >   i. <

2. a. 8,039 < 18,309 < 81,390 < 818,039
   b. 5,020 < 52,000 < 250,000 < 520,000

3. a. 54,000   b. 8,708   c. 11,101   d. 144,000   e. 5,606   f. 8,909

4. a. >   b. <   c. >   d. >   e. >   f. >   g. <   h. >

5. a.

15,000	15,100	15,200	15,300	15,400	15,500	15,600	15,700	15,800	15,900	16,000	16,100

b.

34,500	34,600	34,700	34,800	34,900	35,000	35,100	35,200	35,300	35,400	35,500	35,600

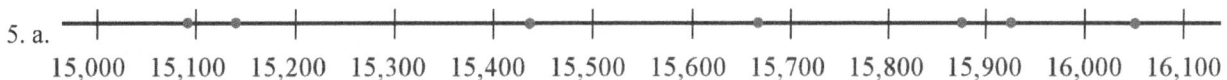

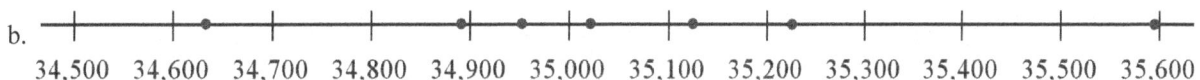

6. 67,030 < 67,049 < 67,250 < 67,370 < 67,510 < 67,703 < 67,780 < 67,940

Comparing with Thousands, continued

7. a. 398,039   b. 290,290   c. 606,660   d. 110,293   e. 301,481   f. 390,200

8. a. 500 < 1,459 < 1,500 < 5,406 < 5,505 < 5,600
   b. 7,800 < 8,708 < 77,988 < 78,707 < 78,777 < 87,600

9. a. $x = 200,000$ b. $x = 70,000$ c. $x = 5,000$

10. a.

n	600	1,200	1,800	2,400	3,000	3,600	4,200
$n + 500$	*1100*	1,700	2,300	2,900	3,500	4,100	4,700

b. It is a skip-counting pattern by 600s, starting at 1,100. The top and bottom numbers both increase by 600. That is because adding 500 to each of the numbers does not change the difference between them.

11. a.

n	52,000	55,000	58,000	61,000	64,000	67,000
$n - 5,000$	*47,000*	50,000	53,000	56,000	59,000	62,000

b. The pattern is: the numbers increase by 3,000 each time.

Adding and Subtracting Big Numbers, p. 65

1. a. 945,601   b. 436,929   c. 818,736   d. 149,259   e. 905,468   f. 524,160

2.

a.	b.	c.
29,100	906,500	610,400
29,300	916,600	610,000
29,500	926,700	609,600
29,700	936,800	609,200
29,900	946,900	608,800
30,100	957,000	608,400
30,300	967,100	608,000
30,500	977,200	607,600
30,700	987,300	607,200
30,900	997,400	606,800

3. a. 85,581   b. 119,976   c. 668,700   d. 66,697   e. 41,893   f. 85,055   g. 426,600   h. 376,935   i. 381,656

4.

a.

419,000 + 1,000   150,000 + 40,000
500 + 36,000   20,000 + 400,000
189,000 + 1,000   36,100 + 400
40,500 + 500   180,000 − 2,000
177,300 + 700   36,000 + 5,000

b.

500,000 − 3,000   140,000 + 70,000
189,000 − 80,000   97,000 + 400,000
40,600 − 500   20,000 + 20,100
250,000 − 40,000   100,000 + 9,000
77,700 − 7,000   100,000 − 29,300

5. a. 302,889   b. 641,571   c. 26,712   d. 876,255

6.

n	13,000	78,000	154,000	500,000	640,500
$n + 1,000$	14,000	79,000	155,000	501,000	641,500
$n + 10,000$	23,000	88,000	164,000	510,000	650,500
$n + 100,000$	113,000	178,000	254,000	600,000	740,500

7. a. 427,443   b. 27,854   c. 25,301   d. 983,715

8. a. >   b. =   c. >   d. =   e. <   f. =

1.

a. 45̲2,550 ≈ 450,000	b. 86̲,256 ≈ 86,000	c. 77,5̲79 ≈ 77,600
d. 245̲,250 ≈ 245,000	e. 8̲94,077 ≈ 900,000	f. 385̲,706 ≈ 386,000
g. 6̲15,493 ≈ 600,000	h. 5̲27,009 ≈ 500,000	i. 2̲52,000 ≈ 300,000
j. 2̲6,566 ≈ 30,000	k. 94̲4,032 ≈ 940,000	l. 335̲,700 ≈ 336,000
m. 48,42̲1 ≈ 48,420	n. 8,5̲55 ≈ 8,600	o. 40̲9,239 ≈ 410,000

2.

a. 10,9̲65 ≈ 11,000	b. 89̲,506 ≈ 90,000	c. 79̲7,329 ≈ 800,000
d. 299̲,850 ≈ 300,000	e. 254,99̲7 ≈ 255,000	f. 599,9̲72 ≈ 600,000

3.

a. 233,5̲64 ≈ 233,600	b. 752̲,493 ≈ 752,000	c. 1̲92,392 ≈ 190,000
d. 89̲5,080 ≈ 900,000	e. 8̲55,429 ≈ 900,000	f. 399̲,477 ≈ 399,000

4.

number	274,302	596,253	709,932	899,430
to the nearest 1,000	274,000	596,000	710,000	899,000
to the nearest 10,000	270,000	600,000	710,000	900,000
to the nearest 100,000	300,000	600,000	700,000	900,000

5. a. about 1,800 days
 b. about 3,300 days
 c. about 3,700 days
 d. about 7,300 days.
 e. In 40 years, you have likely lived 14,610 days, or about 14,600 days.
 f. Answers will vary. For example, if a mother is 36 years old, subtract four times 365 from 14,610.
   And since 365 + 365 = 730, we can subtract two times 730 or 1,460: 14,610 − 1,460 = 13,150.
   This is about 13,200 days.

6. a. 235 ≈ 0
 b. 18,299 ≈ 20,000
 c. 1,392 ≈ 0

7. a. 865 ≈ 1,000
 b. 182 ≈ 0
 c. 5,633 ≈ 6,000

8.

a. 56,250 ≈ 60,000	b. 5,392 ≈ 10,000	c. 2,938 ≈ 0
d. 708,344 ≈ 710,000	e. 599 ≈ 0	f. 44,800 ≈ 40,000

9.

a. That means about <u>236,000</u> people in Purpletown, and about <u>187,000</u> people in
Bluetown. The two towns have approximately <u>423,000</u> people in all.
There are about <u>49,000</u> more people in Purpletown than in Bluetown.

b. There were about <u>3,500</u> live births in total in those two. Seagull hospital had
about <u>1,300</u> more live births than Sunshine hospital.

c. The Nile river (in Africa) is 6,700 km long and the Danube river (in Europe) 2,900 km long.
The Nile is about <u>3,800</u> km longer than the Danube.

c. Rounding to the nearest thousand makes the problem simple. That means about $48,000. So he earns about <u>$4,000</u> monthly.	d. The simplest way is to round the annual mileage to the nearest 10,000, That is about <u>60,000</u> miles. This means he drives about <u>5,000 miles</u> each month.

10. a. See the table.

b. You would need <u>three</u> copies of
Empire State Building to
exceed the height of Burj Khalifa.

c. About 1,000 ft taller.

Building	Height	Height (rounded)
Burj Khalifa	2,717 ft	2,700 ft
Shanghai Tower	2,073 ft	2,100 ft
Taipei 101	1,667 ft	1,700 ft
One World Trade Center	1,776 ft	1,800 ft
Petronas Tower 1	1,483 ft	1,500 ft
Empire State Building	1,250 ft	1,300 ft

11. a. The trip around the equator is about <u>25,000</u> miles.

The Moon is about <u>239,000</u> miles from the Earth.

b. <u>Ten</u> trips around the equator would be a longer distance than the distance from Earth to Moon.

Trips Around The Equator	Approximate Distance (miles)	Trips Around The Equator	Approximate Distance (miles)	Trips Around The Equator	Approximate Distance (miles)
1	25,000	5	125,000	9	225,000
2	50,000	6	150,000	10	250,000
3	75,000	7	175,000	11	275,000
4	100,000	8	200,000	12	300,000

Puzzle corner:

a. Jake's yearly earnings are $47,807.
That means about $ <u>48,000</u> .
So, he earns about <u>$4,000</u> *monthly*.

b. Jack drove 58,496 miles last year.
That is about <u>60,000</u> miles. This means
he drives about <u>5,000 miles</u> each *month*.

1.

a. $11 \times 10 = 110$ $29 \times 10 = 290$	b. $50 \times 10 = 500$ $80 \times 10 = 800$	c. $200 \times 10 = 2,000$ $1,000 \times 10 = 10,000$

2.

a. $7 \times 100 = 700$ $9 \times 100 = 900$	b. $10 \times 100 = 1,000$ $13 \times 100 = 1,300$	c. $20 \times 100 = 2,000$ $22 \times 100 = 2,200$

3.

a. $311 \times 100 = 31,100$ $70 \times 100 = 7,000$ $120 \times 100 = 12,000$	b. $10 \times 19 = 190$ $999 \times 10 = 9,990$ $10 \times 4,500 = 45,000$	c. $60 \times 1,000 = 60,000$ $493 \times 1,000 = 493,000$ $1,000 \times 500 = 500,000$

4.

a. 49 thousands 49,000 49 hundreds 4,900 49 tens 490	b. 20 tens 200 20 hundreds 2,000 20 thousands 20,000	c. 37 tens 370 37 hundreds 3,700 37 thousands 37,000

5.

a. 10 tens <u>100</u> 100 tens <u>1,000</u>	b. 10 hundreds <u>1,000</u> 100 hundreds <u>10,000</u>	c. 100 thousands <u>100,000</u> 1,000 thousands <u>1,000,000</u>

6. a. $4,000   b. $500   c. $20,000

7. a. 1,492   b. $2,500; $1,200   c. 1,900; 1,960

8.

a. $67 \times 10 = 670$ $18 \times 100 = 1,800$ $20 \times 10 = 200$	b. $112 \times 100 = 11,200$ $80 \times 1,000 = 80,000$ $390 \times 10 = 3,900$	c. $44 \times 100 = 4,400$ $90 \times 10 = 900$ $60 \times 1,000 = 60,000$

9.

a. $120 \div 10 = 12$ $600 \div 10 = 60$ $1,300 \div 10 = 130$	b. $700 \div 100 = 7$ $5,600 \div 100 = 56$ $65,000 \div 100 = 650$	c. $12,000 \div 1,000 = 12$ $689,000 \div 1,000 = 689$ $400,000 \div 1,000 = 400$

10.

a.

90	$= 9 \times 10$
100	$= 10 \times 10$
110	$= 11 \times 10$
280	$= 28 \times 10$
1,000	$= 100 \times 10$
4,560	$= 456 \times 10$

b.

900	$= 9 \times 100$
1,000	$= 10 \times 100$
1,100	$= 11 \times 100$
4,000	$= 40 \times 100$
5,900	$= 59 \times 100$
10,000	$= 100 \times 100$

c.

10,000	$= 10 \times 1,000$
15,000	$= 15 \times 1,000$
18,000	$= 18 \times 1,000$
50,000	$= 50 \times 1,000$
160,000	$= 160 \times 1,000$
520,000	$= 520 \times 1,000$

11. Notice something special about these multiples: ALL multiples of 10 end in 0.
ALL multiples of 100 end in 00. ALL multiples of 1000 end in 000.

Multiples of 10, 100, and 1,000, continued

12. Answers vary.

Multiples of ten	What number times 10?
540	$= 54 \times 10$
9,870	$= 987 \times 10$
42,090	$= 4,209 \times 10$

Multiples of hundred	What number times 100?
48,200	$= 482 \times 100$
63,600	$= 636 \times 100$
849,000	$= 8,490 \times 100$

13.

a. $1,000 \div 100 = 10$ $2,100 \div 100 = 21$ $99,900 \div 100 = 999$	b. $90 \div 10 = 9$ $7,000 \div 10 = 700$ $34,800 \div 10 = 3,480$	c. $2,000 \div 1,000 = 2$ $30,000 \div 1,000 = 30$ $342,000 \div 1,000 = 342$

Puzzle corner:
100,000   200,000   300,000

Mixed Review Chapter 2, p. 76

1. Number sentence: $\$23.50 + \$19.90 + \$6.60 = x$
   $x = \$50$
   He paid with a $50-dollar bill.

   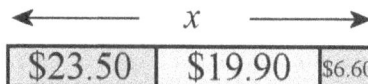

2. a. $56.25   b. $283.01

3. The total cost is $(\$15 - \$3) \times 4$.

4. $5,209 < 25,539 < 25,925 < 525,009$

5. a. $x = 47$ b. $x = 1,266$ c. $x = 633$

6. a. $158 − $38 = $120. Grandma gave him $120.
   b. $3 \times \$4 + \$28 = \$40$. Jill received $40 for her birthday.
   c. $60 − 2 \times \$11 = \$38$. He had $38 left.
   d. $3 \times \$0.60 + \$0.80 = \$2.60$. The total cost was $2.60.
      He received $10 − $2.60 = $7.40 in change.

7. a. $60,000 + 8,000 + 50 + 6$
   b. $800,000 + 10,000 + 5,000 + 200 + 20 + 4$

8. a. $1.00   b. $8.00   c. $35.00   d. $166.00
   e. $95.00   f. $99.00   g. $100.00   h. $101.00

9. a. $20 + $15 + $25 = $60.
   b. $4,000 \times \$1 + 1,000 \times \$1 = \$5,000$

1. a. 13,094   b. 306,050   c. 1,000,000

2. a. 785,300   b. 70,008

3. a. three thousand   b. thirty   c. 300 thousand   d. 30 thousand

4.

n	78,974	5,367	2,558	407,409	299,603
rounded to nearest 1,000	79,000	5,000	3,000	407,000	300,000
rounded to nearest 10,000	80,000	10,000	0	410,000	300,000

5. Estimate: $5,100 - 2,800 - 700 = 1,600$. Exact: 1,556.

6. a. 500   b. 700,000   c. 600,000

7. a. $5,406 < 5,604$
   b. $49,530 < 49,553$
   c. $605,748 > 60,584$

8. $95,695 < 145,900 < 495,644 < 496,455 < 590,554 < 5,905,544$

9. a. 392,054
   b. 444,869

10. $500 \times \$100 = \$50,000$

11. In ten months, Mark earns $10 \times \$2,560 = \underline{\$25,600}$.
    In two months, he earns $\$2,560 + \$2,560 = \underline{\$5,120}$.
    In 12 months, he earns $\$25,600 + \$5,120 = \underline{\$30,720}$.

Chapter 3: Multiplication

Understanding Multiplication, p. 84

1. a. $2 + 2 + 2 + 2 = 4 \times 2 = 8$; $20 + 20 + 20 + 20 = 4 \times 20 = 80$
   b. $8 + 8 + 8 = 3 \times 8 = 24$; $80 + 80 + 80 = 3 \times 80 = 240$;
   c. $500 + 500 + 500 + 500 = 4 \times 500 = 2,000$ $120 + 120 + 120 = 3 \times 120 = 360$

2. a. $6 \times 3 = 18$; $3 \times 6 = 18$ b. $5 \times 1 = 5$; $1 \times 5 = 5$

3. a. 16, 0   b. 15, 10   c. 16, 8   d. 30, 27

4. a. 48, 0, 16   b. 300; 6,000; 12,000   c. 4,000; 1,000; 633   d. 68, 63, 200

5. a. 6, 0   b. 200, 250   c. 3, 81

6. a. factors; product   b. $4 \times 8 = 32$ c. The product is 0.   d. 5, because $2 \times 6 \times \underline{5} = 60$

7. a. $3 \times 12 + 5 = 41$ eggs in all.   b. Jack had: $6 \times 10 - 3 = 57$ left.
   c. $4 \times 10 + 3 \times 6 = 58$ crayons in all.   d. Ernest's change was: $\$50 - 3 \times \$11 = \$17$.
   e. $5 \times 3 + 7 \times 2 = 29$ wheels

8.

a. $7 \times 10 = \underline{?}$
    $\underline{? = 70}$
	b. $\underline{?} \times 4 = 24$
	    $\underline{? = 6}$

c. $\underline{?} \times 2 = 18$
    $\underline{? = 9}$
	d. $y \times 4 = 36$
	    $y = 9$

e. $y \times 10 = 150$
    $y = 15$
	f. $y \times 12 = 60$
	    $y = 5$

g. $4 \times 20 = y$
    $y = 80$
	h. $y \times 7 = 35$
	    $y = 5$

i. $300 \times y = 1,200$
    $y = 4$
	j. $y \times 2 = 40$
	    $y = 20$

Multiplication Tables Review, p. 87

1.

$1 \times 5 = 5$	$7 \times 5 = 35$	$1 \times 10 = 10$	$7 \times 10 = 70$	$1 \times 11 = 11$	$7 \times 11 = 77$
$2 \times 5 = 10$	$8 \times 5 = 40$	$2 \times 10 = 20$	$8 \times 10 = 80$	$2 \times 11 = 22$	$8 \times 11 = 88$
$3 \times 5 = 15$	$9 \times 5 = 45$	$3 \times 10 = 30$	$9 \times 10 = 90$	$3 \times 11 = 33$	$9 \times 11 = 99$
$4 \times 5 = 20$	$10 \times 5 = 50$	$4 \times 10 = 40$	$10 \times 10 = 100$	$4 \times 11 = 44$	$10 \times 11 = 110$
$5 \times 5 = 25$	$11 \times 5 = 55$	$5 \times 10 = 50$	$11 \times 10 = 110$	$5 \times 11 = 55$	$11 \times 11 = 121$
$6 \times 5 = 30$	$12 \times 5 = 60$	$6 \times 10 = 60$	$12 \times 10 = 120$	$6 \times 11 = 66$	$12 \times 11 = 132$

2. The products (answers) 10, 20, 30, 40, 50, and 60 are found both in the table of 5 and in the table of 10. That is because $10 = 2 \times 5$, so any product of 10 also is a product of 5.

$1 \times 2 = 2$	$7 \times 2 = 14$	$1 \times 4 = 4$	$7 \times 4 = 28$	$1 \times 8 = 8$	$7 \times 8 = 56$
$2 \times 2 = 4$	$8 \times 2 = 16$	$2 \times 4 = 8$	$8 \times 4 = 32$	$2 \times 8 = 16$	$8 \times 8 = 64$
$3 \times 2 = 6$	$9 \times 2 = 18$	$3 \times 4 = 12$	$9 \times 4 = 36$	$3 \times 8 = 24$	$9 \times 8 = 72$
$4 \times 2 = 8$	$10 \times 2 = 20$	$4 \times 4 = 16$	$10 \times 4 = 40$	$4 \times 8 = 32$	$10 \times 8 = 80$
$5 \times 2 = 10$	$11 \times 2 = 22$	$5 \times 4 = 20$	$11 \times 4 = 44$	$5 \times 8 = 40$	$11 \times 8 = 88$
$6 \times 2 = 12$	$12 \times 2 = 24$	$6 \times 4 = 24$	$12 \times 4 = 48$	$6 \times 8 = 48$	$12 \times 8 = 96$

The products (answers) 8, 16, and 24 are in the tables of 2, 4, and 8. Because $8 = 2 \times 4$, any product of 8 also is a product of 2 and of 4. For example, $24 = 3 \times 8 = 6 \times 4 = 12 \times 2$.

3.

$1 \times 3 = 3$ $7 \times 3 = 21$ $2 \times 3 = 6$ $8 \times 3 = 24$ $3 \times 3 = 9$ $9 \times 3 = 27$ $4 \times 3 = 12$ $10 \times 3 = 30$ $5 \times 3 = 15$ $11 \times 3 = 33$ $6 \times 3 = 18$ $12 \times 3 = 36$	$1 \times 6 = 6$ $7 \times 6 = 42$ $2 \times 6 = 12$ $8 \times 6 = 48$ $3 \times 6 = 18$ $9 \times 6 = 54$ $4 \times 6 = 24$ $10 \times 6 = 60$ $5 \times 6 = 30$ $11 \times 6 = 66$ $6 \times 6 = 36$ $12 \times 6 = 72$	$1 \times 9 = 09$ $7 \times 9 = 63$ $2 \times 9 = 18$ $8 \times 9 = 72$ $3 \times 9 = 27$ $9 \times 9 = 81$ $4 \times 9 = 36$ $10 \times 9 = 90$ $5 \times 9 = 45$ $11 \times 9 = 99$ $6 \times 9 = 54$ $12 \times 9 = 108$	

The products (answers) 6, 12, 18, 24, 30, and 36 are in tables of 3 and 6. Because $6 = 2 \times 3$, any product of 6 is also a product of 3. For example, $30 = 5 \times 6 = 10 \times 3$.

In the table of nine, the tens digits of the products (colored red) go from 0 to 9. Then there is a 9 (of 99) and then a 10 (of 108). From then on they would once again continue in order. The ones digits start at 9 and decrease one by one to 0, then start over again.

4.

$1 \times 7 = 7$ $7 \times 7 = 49$ $2 \times 7 = 14$ $8 \times 7 = 56$ $3 \times 7 = 21$ $9 \times 7 = 63$ $4 \times 7 = 28$ $10 \times 7 = 70$ $5 \times 7 = 35$ $11 \times 7 = 77$ $6 \times 7 = 42$ $12 \times 7 = 84$	$1 \times 12 = 12$ $7 \times 12 = 84$ $2 \times 12 = 24$ $8 \times 12 = 96$ $3 \times 12 = 36$ $9 \times 12 = 108$ $4 \times 12 = 48$ $10 \times 12 = 120$ $5 \times 12 = 60$ $11 \times 12 = 132$ $6 \times 12 = 72$ $12 \times 12 = 144$

5. a. 7, 4, 8   b. 8, 5, 9   c. 8, 6, 7   d. 8, 6, 7   e. 9, 7, 8   f. 9, 6, 7
   g. 7, 9, 8   h. 9, 8, 6   i. 5, 9, 3   j. 12, 5, 6   k. 6, 12, 7   l. 9, 2, 4

6.

×	0	1	2	3	4	5	6	7	8	9	10	11	12
0	0	0	0	0	0	0	0	0	0	0	0	0	0
1	0	1	2	3	4	5	6	7	8	9	10	11	12
2	0	2	4	6	8	10	12	14	16	18	20	22	24
3	0	3	6	9	12	15	18	21	24	27	30	33	36
4	0	4	8	12	16	20	24	28	32	36	40	44	48
5	0	5	10	15	20	25	30	35	40	45	50	55	60
6	0	6	12	18	24	30	36	42	48	54	60	66	72
7	0	7	14	21	28	35	42	49	56	63	70	77	84
8	0	8	16	24	32	40	48	56	64	72	80	88	96
9	0	9	18	27	36	45	54	63	72	81	90	99	108
10	0	10	20	30	40	50	60	70	80	90	100	110	120
11	0	11	22	33	44	55	66	77	88	99	110	121	132
12	0	12	24	36	48	60	72	84	96	108	120	132	144

7. $90 - 8 \times 7 = 34$. So, <u>34 students went by bus</u>.

8. a. 33   b. 26   c. 8   d. 70   e. 63   f. 26

9.

	a. $2 \times 6 = 4 \times \underline{3}$	b. $6 \times 6 = 4 \times \underline{9}$
c. $3 \times 10 = 6 \times \underline{5}$	d. $2 \times 20 = 10 \times \underline{4}$	e. $5 \times 12 = 6 \times \underline{10}$

Scales Problems, p. 90

1. a. 7   b. 6   c. 6   d. 2   e. 7   f. 4   g. 2   h. 7

2. a. 11   b. 8   c. 13   d. 5

3. a. Hint: from the first balance we learn that one circle + one rectangle weigh 14. In the second balance, we have two circles and one rectangle, and of those, one rectangle and one circle together still weigh 14. This means that the one circle must weigh 6.
   Solution: 1 rectangle weighs 8 and 1 circle weighs 6.

   b. Hint: guess and check using small numbers. Look at the second scales, and in it, try for example one circle being 2. Then you will notice that that will not work! The circle has to be a bigger number. Check to see if the circle is 4. Then the diamond would be 0. That will not work either. Check to see if the circle is 6. Then the diamond would be 4. Now go to the first scale and check, if the circle = 6 and the diamond = 4 works there. If not, change your guess.
   Solution: 1 circle weighs 5 and 1 diamond weighs 2.

4. a. 1 circle weighs 4 and 1 square weighs 17.   b. 1 square weighs 5 and 1 triangle weighs 3.
   c. 1 square weighs 3 and 1 circle weighs 11.   d. 1 circle weighs 2 and 1 triangle weighs 3.

5. a. 70   b. 29   c. 60   d. 70   e. 9   f. 21   g. 18   h. 17   i. 8

Multiplying by Whole Tens and Hundreds, p. 94

1. a. 3,150; 35,600; 3,500   b. 620,000; 12,000; 13,000   c. 250,000; 38,000; 50,000

2. a. 160; 80; 100   b. 1400; 1,000; 2,200   c. 240; 700; 1,800   d. 320; 8,400; 1,080

3.

a. 20×7	b. 20×5	c. 200×8	d. 200×25
$= \underline{10} \times 2 \times 7$	$= 10 \times 2 \times 5$	$= 100 \times 2 \times 8$	$= 100 \times 2 \times 25$
$= 10 \times \underline{14}$	$= 10 \times 10$	$= 100 \times 16$	$= 100 \times 50$
$= \underline{140}$	$= 100$	$= 1600$	$= 5000$

4. $A = 20 \text{ ft} \times 15 \text{ ft} = 300 \text{ ft}^2$

5. $A = 15 \text{ ft} \times 200 \text{ ft} = 3,000 \text{ ft}^2$

6. One truckload costs $5 \times \$20 + \$30 = \$130$. Four truckloads cost $4 \times \$130 = \$520$.

7. a. 120; 160   b. 420; 550   c. 720; 450   d. 660; 480   e. 1,800; 2,800
   f. 4,200; 6,600   g. 2,400; 4,500   h. 3,300; 7,200   i. 1,320; 2400

8. a. 1,800; 21,000   b. 4,800; 27,000   c. 20,000; 40,000   d. 64,000; 100,000   e. 10,000; 1,200   f. 240,000; 99,000

9. One hour has _60_ minutes.
   How many minutes are in 12 hours? $\underline{12 \times 60 \text{ min} = 720 \text{ minutes}}$
   How many minutes are in 24 hours? $\underline{2 \times 720 \text{ min} = 1,440 \text{ minutes (double the previous result)}}$

10. One hour has _60_ minutes, and one minute has _60_ seconds.
    How many seconds are there in one hour? $\underline{60 \times 60 \text{ sec} = 3,600 \text{ seconds}}$

11. a. $\underline{8 \times \$30 = \$240}$ b. $\underline{40 \times \$30 = \$1,200}$

    c. Guess and check: $\underline{2 \times \$240 = \$480; 4 \times \$240 = \$960; 5 \times \$240 = \$1,200; \text{ so five days.}}$

12. a. 120; 9   b. 8; 120   c. 10; 90   d. 160; 9   e. 50; 700   f. 70; 600

Puzzle Corner: $600 \times 50 = 6 \times 100 \times 5 \times 10 = 100 \times 10 \times (6 \times 5) = 1000 \times 30 = 30,000$.

1.

a. 6 × 27 (20 + 7) 6 × <u>20</u> and 6 × <u>7</u> <u>120</u> and <u>42</u> = <u>162</u>	b. 5 × 83 (80 + 3) 5 × <u>80</u> and 5 × 3 <u>400</u> and <u>15</u> = <u>415</u>	c. 9 × 34 (30 + 4) 9 × <u>30</u> and 9 × <u>4</u> <u>270</u> and <u>36</u> = <u>306</u>
d. 3 × 99 3 × <u>90</u> and 3 × <u>9</u> <u>270</u> and <u>27</u> = <u>297</u>	e. 7 × 65 7 × <u>60</u> and 7 × <u>5</u> <u>420</u> and <u>35</u> = <u>455</u>	f. 4 × 58 4 × <u>50</u> and 4 × <u>8</u> <u>200</u> and <u>32</u> = <u>232</u>

2.

a. $9 \times 17 = 9 \times 10 + 9 \times 7$

$= 90 + 63 = 153$

b. $6 \times 29 = 6 \times 20 + 6 \times 9$

$= 120 + 54 = 174$

c. $7 \times 33 = 7 \times 30 + 7 \times 3$

$= 210 + 21 = 231$

3.

a. $7 \times 16 = 7 \times 10 + 7 \times 6$

$= 70 + 42 = 112$

b. $5 \times 21 = 5 \times 20 + 5 \times 1$

$= 100 + 5 = 105$

c. $8 \times 34 = 8 \times 30 + 8 \times 4$

$= 240 + 32 = 272$

Multiply in Parts 1, continued

4.

a. 6 × 19	b. 3 × 73	c. 4 × 67

6 × 10 →   6 0
 6 × 9   →   + 5 4
 	
         1 1 4
	3 × 70 → 2 1 0
	 3 × 3 → + 9
	
	         2 1 9
		4 × 60 → 2 4 0
		 4 × 7 → + 2 8
		
		         2 6 8

d. 5 × 92	e. 9 × 33	f. 7 × 47

5 × 90 →   4 5 0
 5 × 2   →   +   1 0
 	
         4 6 0
	9 × 30 → 2 7 0
	 9 × 3 → + 2 7
	
	         2 9 7
		7 × 40 → 2 8 0
		 7 × 7 → + 4 9
		
		         3 2 9

5. a. 65   b. 135   c. 165   d. 168   e. 88   f. 357

6. a. >   b. >   c. >

7. a. Jack's total cost was $8 \times \$14 = 8 \times \$10 + 8 \times \$4 = \$80 + \$32 = \underline{\$112}$.
   b. There were $9 \times 14 + 56 = \underline{182}$ seats in all.
   c. The cost of the hammer is $3 \times \$17 = \underline{\$51}$.

Multiply in Parts 2, p. 101

1.

a. 3 × 127 (100 + 20 + 7)	b. 5 × 243 (200 + 40 + 3)
3 × <u>100</u> and 3 × <u>20</u> and 3 × <u>7</u> <u>300</u> and <u>60</u> and <u>21</u> = <u>381</u>	5 × <u>200</u> and 5 × <u>40</u> and 5 × <u>3</u> <u>1,000</u> and <u>200</u> and <u>15</u> = <u>1,215</u>

d. 4 × 6,507
   (6000 + 500 + 7)

4 × <u>6,000</u> and 4 × <u>500</u> and 4 × <u>7</u>

 <u>24,000</u> and <u>2,000</u> and <u>28</u>

= <u>26,028</u>

e. 5 × 4,813

5 × 4,000 + 5 × <u>800</u> + 5 × 10 + 5 × <u>3</u>

 <u>20,000</u> + <u>4,000</u> + <u>50</u> + <u>15</u>

= <u>24,065</u>

Multiply in Parts 2, continued

2.

a. <u>4 × 128</u>
4 × 100 →   4 0 0
4 × 20   →   8 0
4 × 8   →   +   3 2
                            —————
                         5 1 2
	b. <u>8 × 151</u>
	
	      8 0 0
	      4 0 0
	+ 8
	—————
	   1 2 0 8
		c. <u>3 × 452</u>
		
		   1 2 0 0
		     1 5 0
		+ 6
		—————
		   1 3 5 6

d. <u>6 × 3,217</u>
 	
   1 8 0 0 0
     1 2 0 0
       6 0 0
+   4 2
—————
   1 9 3 0 2
	e. <u>8 × 2,552</u>
	
	   1 6 0 0 0
	     4 0 0 0
	       4 0 0
	+ 1 6
	—————
	   2 0 4 1 6
		f. <u>6 × 1,098</u>
		
		    6 0 0 0
		          0
		      5 4 0
		+ 4 8
		—————
		    6 5 8 8

3. a. In half a year, Dad pays 6 × \$138 = \$600 + \$180 + \$48 = <u>\$828</u>.

   b. The perimeter is 4 × 255 cm = 800 cm + 200 cm + 20 cm = <u>1,020 cm</u>.

   c. The bigger roll contains 5 × 56 cm = 250 cm + 30 cm = 280 cm of material.
     In total the rolls have 56 cm + 280 cm = <u>336 cm</u> of material.

Multiply in Parts – Area Model, p. 103

1.

a.   8 × 127
    = 8 × 100 + 8 × 20 + 8 × 7
    The total area is <u>1,016</u> square units.	

b.   6 × 245
    = 6 × 200 + 6 × 40 + 6 × 5
    The total area is <u>1,470</u> square units	

c.   9 × 196
    = 9 × 100 + 9 × 90 + 9 × 6
    The total area is <u>1,764</u> square units.	

2. a. 6 × 6 = 9 × <u>4</u>   b. <u>12</u> × 10 = 5 × 24

   c. 20 + <u>20</u> = 4 × 10   d. 6000 = 30 × <u>200</u>

   e. 120 − 75 = 5 × <u>9</u>   f. <u>750</u> + 750 = 5 × 300

Multiply in Parts — Area Model, cont.

3.

a. 7 × 153 Areas of the parts: 700 + 350 + 21 Total area: 1,071	100 50 3 7
b. 5 × 218 Areas of the parts: 1,000 + 50 + 40 Total area: 1,090	200 10 8 5
c. 8 × 376 Areas of the parts: 2,400 + 560 + 48 Total area: 3,008	300 70 6 8

4. a. Susie orders 5 × 72 = <u>360</u> flowers in 5 weeks.
   b. It costs her 5 × $70 = <u>$350</u> for five weeks of orders.

Multiplying Money Amounts, p. 105

1.

a. 6 × $11.85	b. 5 × $2.93
6 × $11 → $6 6.0 0 6 × $0.80 → 4.8 0 6 × $0.05 → + 0.3 0 ─────── $7 1.1 0	5 × $2 → $1 0.0 0 5 × $0.90 → 4.5 0 5 × $0.03 → + 0.1 5 ─────── $1 4.6 5
c. 7 × $3.75	d. 8 × $10.95
7 × $3 → $2 1.0 0 7 × $0.70 → 4.9 0 7 × $0.05 → + 0.3 5 ─────── $2 6.2 5	8 × $10 → $8 0.0 0 8 × $0.90 → 7.2 0 8 × $0.05 → + 0.4 0 ─────── $8 7.6 0

2.

a. 6 × $2.80	b. 5 × $4.70
<u>$12</u> + <u>$4.80</u> = <u>$16.80</u> (6 × $2) (6 × $0.80)	<u>$20</u> + <u>$3.50</u> = <u>$23.50</u> (5 × $4) (5 × $0.70)
c. 4 × $12.50	d. 7 × $5.61
$48 + $2 = $50	$35 + $4.20 + $0.07 = $39.27

3. a. Her total cost was 5 × $2.00 + 5 × $0.70 = $10.00 + $3.50 = <u>$13.50</u>.
   b. Her change was $20.00 − $13.50 = <u>$6.50</u>.

4. a. The cost for four is 4 × 23.50 = 4 × $20 + 4 × $3 + 4 × $0.50 = $80 + $12 + $2 = <u>$94</u>.
   b. Their change is $100.00 − $94.00 = <u>$6.00</u>.

Multiplying Money Amounts, cont.

5.

a. Start at 80. Add 40 each time:	b. Start at 42,000. Subtract 3,000 each time:	c. Start at 1. Add 5 each time:
_120_	42,000	1
160	39,000	6
200	36,000	11
240	33,000	16
280	30,000	21
320	27,000	26
360	24,000	31
		36
What does this pattern remind you of? The multiplication table of 4.	What does this pattern remind you of? The multiplication table of 3.	41

6. When you add 5 to a number, it changes parity (goes from even to odd, or from odd to even). In other words, if you add 5 to an odd number, you get an even number, and vice versa. In fact, the same happens when you repeatedly add any odd number.

Estimating in Multiplication, p. 107

1. Answers may vary. Estimating is not an "exact science".
   a. $5 \times 70 = 350$ b. $11 \times 60 = 660$ c. $120 \times 8 = 960$ d. $30 \times 50 = 1,500$ e. $7 \times \$4 = \$28$
   f. $8 \times \$12 = \$96$ g. $25 \times \$40 = \$1,000$ h. $9 \times 20 = 180$ or $10 \times 17 = 170$ i. $60 \times 900 = 54,000$

2. a. $20 \times \$45 = \$900$; however since this is about shopping, it might be better not to round down so much, and estimate the cost as $24 \times \$45$, which you can calculate in two parts: $20 \times \$45$ and $4 \times \$45$, which is $\$900 + \$180 = \$1,080$.
   b. $500 \times 20\cent = 10,000\cent = \$100$
   c. $200 \times \$1.50 = \$200 + \$100 = \$300$
   d. Tennis balls: $6 \times \$3 = \$18$; Rackets: $2 \times \$12 = \$24$; Total cost: $\$42$.

3. a. Bill can buy 5 ads. Round to $350. Then add: two ads is $700, four ads is $1,400, six ads is $2,100. So he can not afford six ads. Five ads would be $1,400 + $350 = $1,750.
   b. Round the rate to $3 per hour. Since $8 \times \$3 = \$24$, she can rent them for about 8 hours.
   c. Beans: $8 \times \$0.30 = \$2.40$; Lentils: $5 \times \$0.40 = \$2$. So it is cheaper to buy 5 bags of lentils.
   d. Round the cost of string to $0.20 per foot. Round the number of children to 30. Cost per one child is about $8 \times \$0.20 = \$1.60$. The total cost is about $30 \times \$1.60 = \$48$.

Multiply in Columns—the Easy Way, p. 109

1. a. 456   b. 445   c. 301   d. 312   e. 415   f. 564   g. 288   h. 287

2. a. 822   b. 872   c. 2,191   d. 2,256   e. 3,608   f. 2,085
   g. 6,240   h. 1,944   i. 5,562   j. 1,698   k. 2,056   l. 3,040

3. a. 6   b. 90   c. 90

4. a. 581   b. 565

5. a. $(236 − \$40) \times 7 = \$196 \times 7 = \$1,372$
   b. 992 ft

6. a. 40   b. 40   c. 700
   d. 12   e. 100   f. 80

Puzzle corner. a. 172; 1204   b. 358; 32   c. 709; 00

Multiply in Columns—the Easy Way, part 2, p. 112

1. a. 10,628   b. 30,528   c. 24,318   d. 56,712   e. 11,212   f. 26,460

2. a. 9,930   b. 18,495   c. 22,960   d. 36,702   e. 12,445   f. 34,408

3. a. The total distance is $2 \times 2 \times 3{,}820 = 4 \times 3{,}820 = 15{,}280$ feet.
   b. $55 + 4 \times 55 = 275$ marbles in total.

4. a. $5.49   b. $35.48   c. $80.50   d. $61.50
   e. $60.30   f. $58.94   g. $82.60   h. $318.48

5. a. Estimate: $10 \times 1.57 = 15.70$ or $10 \times 1.60 = 16.00$. Total cost: $9 \times \$1.57 = \$14.13$
   b. Estimate: $8 \times 2.3 = 18.40$. Change: $\$20 - 8 \times \$2.28 = \$20 - \$18.24 = \$1.76$

Multiply in Columns, the Standard Way, p. 115

1.

a.
```
  2              53
  53         x    8
x  8        ─────────
─────           24
4 2 4          400
              ─────
               424
```

b.
```
  2             88
  88         x   3
x  3        ─────────
─────          24
2 6 4         240
             ─────
              264
```

2.

a.
```
  2             79
  79         x   3
x  3        ─────────
─────          27
2 3 7         210
             ─────
              237
```

b.
```
  4             18
  18         x   5
x  5        ─────────
─────          40
 9 0           50
             ─────
              90
```

3. a. 306   b. 57   c. 124   d. 322   e. 396   f. 351
   g. 261   h. 134   i. 180   j. 432   k. 204   l. 92

4. a. Three chairs: $3 \times \$48 = \$144$. Six chairs: $2 \times \$144 = \$288$ (double the previous result).
   b. In five days, you earn $5 \times \$77 = \$385$.
      In ten days, you earn $10 \times \$77 = \$770$.

5.

a.
```
  123            123
x   8        x     8
─────        ─────────
 984            24
               160
               800
             ─────
              984
```

b.
```
  2 2           279
  279        x     3
x   3        ─────────
─────           27
 837           210
               600
             ─────
              837
```

c.
```
  463            463
x   5        x     5
─────        ─────────
2 3 1 5         15
               300
              2000
             ─────
              2315
```

d.
```
  156            156
x   6        x     6
─────        ─────────
 936            36
               300
               600
             ─────
              936
```

6. a. 924   b. 3,542   c. 1,390   d. 2,233   e. 864   f. 7,281
   g. 861   h. 734   i. 10,872   j. 7,542   k. 24,708   l. 47,970

7. Perimeter: 2×9 ft $+ 2 \times 28$ ft $= 18$ ft $+ 56$ ft $= 74$ ft.
   Area: 9 ft $\times 28$ ft $= 252$ sq. ft.

8. a. Every time she multiplies she carries the ones digit instead of the tens digit.
      The real answers are: 234, 342, 532.
   b. Andy does not carry. He writes the tens digit that he should carry as part of the answer.
      The real answers are: 84, 225, 645.

Multiplying in Columns, Practice, p. 119

1. a. 387
   b. Estimation: $8 \times 70 = 560$; Exact: 576
   c. Estimation: $3 \times 770 = 2,310$; Exact: 2,313
   d. Estimation: $5 \times 800 = 4,000$; Exact: 4,095
   e. Estimation: $4 \times 2,500 = 10,000$; Exact: 10,084
   f. Estimation: $3 \times 9,000 = 27,000$; Exact: 26,136

2. It has $3 \times 187 = 561$ pages.

3. Five packages have $5 \times 250 = 1,250$ sheets, which is not enough. $6 \times 250 = 1,500$, which is enough.
   So, she needs to buy <u>six</u> packages.

4. a. Four buses could seat $4 \times 43 = 172$ students.
   b. Seven buses could seat $7 \times 43 = 301$ students.
   c. They need eight buses.

5. a. 80   b. 3   c. 30

6. a. Estimation: $5 \times 200 = 1,000$; Exact: 980
   b. Estimation: $9 \times 200 = 1,800$; Exact: 1,845
   c. Estimation: $9 \times 10,000 = 90,000$; Exact: 88,263
   d. Estimation: $6 \times 5,000 = 30,000$; Exact: 28,860

Puzzle Corner: $117 \times 4 = 468$; $174 \times 5 = 870$; $138 \times 7 = 966$; $3,219 \times 3 = 9657$

Order of Operations Again, p. 121

1. a. First calculate the sum <u>23 + 31</u>.
   b. Next multiply that sum by <u>9</u>.
   c. Subtract that result from <u>650</u>.
   d. Lastly add <u>211</u> to the subtraction result.
      Result: 375

2. a. 1,800; 150   b. 210; 2,000   c. 3,200; 0   d. 2,800; 5,600

3. a. 100; 900   b. 3,500; 700   c. 600; 1,400

4. a. 100   b. 900
   c. 5,800   d. 230
   e. 860   f. 2,700

5. a. 316   b. 2,155   c. 1,124

6. a. $4 \times \$2 + 3 \times \$3 = \$17$
   b. $4 \times (\$2 + \$3) = \$20$
   c. $\$50 - 5 \times \$3 - 5 \times \$2 = \$25$

7. a. $N = 3$ b. $N = 12$ c. $N = 3$
   d. $N = 1,500$ e. $N = 9$ f. $N = 4$

8. a. Her change was $\$200 - 7 \times \$25 = \underline{\$25}$.
   b. The total weight is $8 \times 3 \text{ kg} + 15 \times 2 \text{ kg} = \underline{54 \text{ kg}}$.
   c. The second building is $3 \times 9 \times 9 \text{ ft} = \underline{243 \text{ feet tall}}$.
   d. $2 \times \$18 - 9 \times \$3 = \$9$. The nine cans of cat food is \$9 cheaper than two boxes of cat food.

Puzzle Corner: a. $7 \times (2 + 8) = 70$ b. $80 - 5 \times (10 - 5) = 55$ c. $(4 + 8) \times 5 - 20 = 40$

1. a. Estimation: 4 × $5 = $20; answer $18.20
   b. Estimation: 4 × $10 = $40; answer $38.80
   c. Estimation: 7 × $5 = $35; answer $34.37
   d. Estimation: 6 × $1= $6; answer $4.92
   e. Estimation: 7 × $13 = $91; answer $87.71
   f. Estimation: 5 × $43 = $215; answer $215.75

2. Estimates vary. For example: 8 × $1.40 = $11.20. Change: $8.80.
   Solution: $8.88. Multiply in columns or in parts 8 × $1.39 = $11.12. Then, subtract $20 − $11.12 = $8.88.

3.

---

a.

|←———————— $50 ————————→|

| $7.20 | $7.20 | $7.20 | x |

3 × $7.20 + x = $50

AND

x = $50 − 3 × $7.20
x = $28.40

---

b.

|←———————— x ————————→|

| $29 | $29 | $29 | $29 | $51 |

x = 4 × $29 + $51
x = $167

---

c.

|←———————— $30 ————————→|

| $3.08 | $3.08 | $3.08 | $3.08 | $3.08 | x |

5 × $3.08 + x = $30

AND

x = $30 − 5 × $3.08

x = $14.60

---

d.

|←———————— x ————————→|

| $11.50 | $11.50 | $11.50 | $11.50 | $11.50 | $12.50 |

x = 5 × $11.50 + $12.50

x = $70

---

4. Notice that you need <u>five</u> packs because five packs will contain 20 bottles.
   Estimation: 5 × $3 = $15.
   Calculation: 5 × $2.76 = $13.80.

5. Estimation: 20 × $0.15 = 300 cents = $3; 10 × $1 = $10; Total $13.
   Calculation: 20 × $0.15 = 300 cents = $3. The student can solve 10 × $1.09 by multiplying in parts: 10 × $1
   is $10, and 10 × 9 cents = 90 cents. The notebooks cost $10.90, and the total cost is $13.90.

6. Estimation: 4 × $10 + $65 + $26 = $40 + $65 + $26 = $131.
   Calculations: 4 × $9.80 = $39.20. $39.20 + $65 + $25.80 = $130

Puzzle corner. $20. The price of one wheelbarrow was $125 because 8 × $125 = $1000.

1. a.

Miles	45	90	135	180	225	270	315	360	405	450
Hours	1	2	3	4	5	6	7	8	9	10

b.

Dollars	$5.10	$10.20	$15.30	$20.40	$25.50	$30.60	$35.70	$40.80	$45.90	$51.00
Meters	1	2	3	4	5	6	7	8	9	10

c.

Dollars	$1.50	$3.00	$4.50	$6.00	$7.50	$9.00	$10.50	$12.00	$13.50	$15.00
Cans	1	2	3	4	5	6	7	8	9	10

d.

Dollars	$30	$60	$90	$120	$150	$180	$210	$240	$270	$300
Buckets	2	4	6	8	10	12	14	16	18	20

e.

Feet	40	80	120	160	200	240	280	320	360	400
Minutes	10	20	30	40	50	60	70	80	90	100

f.

Tires	Minutes
1	15
2	30
3	45
4	60
5	75
6	90

g.

Days	Scarves
3	1
6	2
9	3
12	4
15	5
18	6

h.

Hours	Dollars
1	$15
2	$30
3	$45
4	$60
5	$75
6	$90

Two identical bags contain 12 kg of potatoes in total. How many bags containing 12 kg would you need to have 30 kg of potatoes?

To solve this problem, you can make a chart as the one on the right:

1 bag	<u>6</u> kg
2 bags	12 kg
3 bags	3 × 6 kg = 18 kg
<u>5</u> bags	<u>5</u> × 6 kg = 30 kg

The total number of chocolates in five identical boxes was 30. How many chocolates would two boxes contain?

First find out how many in ONE box:

1 box	<u>6</u> chocolates
2 boxes	<u>12</u> chocolates
5 boxes	30 chocolates

So Many of the Same Thing, continued

2. a.

1 flower	$3
5 flowers	$15
6 flowers	$18

b.

1 can	200 g
3 cans	600 g
4 cans	800 g

c.

1 lure	$2
3 lures	$6
7 lures	$14

d.

1 episode	30 min
3 episodes	90 min
5 episodes	150 min

e.

1 sit-up	2 sec
5 sit-ups	10 sec
30 sit-ups	60 sec

f.

1 notebook	$2
7 notebooks	$14
10 notebooks	$20

g.

1 day	30 pages
4 days	120 pages
10 days	300 pages

h.

1 pair	$0.75
6 pairs	$4.50
30 pairs	$22.50

3. a.

5 cars	$35.50
1 car	$7.10
4 cars	$28.40

b.

5 rows	100 minutes
1 row	20 minutes
9 rows	180 minutes = 3 hours

c. 25 minutes.

Solution: Elaine can run 4 times around a track in an hour. This means she takes 15 minutes to run around the track. Today she ran three times. She took 45 minutes for that. Then walked the fourth time. All that took 10 minutes longer than on her normal days. This means she took 1 hour 10 minutes. So, if running took 45 minutes, and in total she used 1 h 10 min, then the walking time is the difference of those, which is 25 minutes.

Multiplying Two-Digit Numbers in Parts, p. 130

1. a. $23 \times 31 = 20 \times 30 + 20 \times 1 + 3 \times 30 + 3 \times 1 = 600 + 20 + 90 + 3 = 713$ square units.
   b. $28 \times 45 = 20 \times 40 + 20 \times 5 + 8 \times 40 + 8 \times 5 = 800 + 100 + 320 + 40 = 1{,}260$ square units.
   c. $35 \times 27 = 30 \times 20 + 30 \times 7 + 5 \times 20 + 5 \times 7 = 600 + 210 + 100 + 35 = 945$ square units.

2.

a. $13 \times 27 =$

$10 \times 20 + 10 \times 7$

$+ 3 \times 20 + 3 \times 7$

$= 200 + 70 + 60 + 21 = 351$

b. $36 \times 25 =$

$30 \times 20 + 30 \times 5$

$+ 6 \times 20 + 6 \times 5$

$= 600 + 150 + 120 + 30 = 900$

c. $28 \times 49 =$

$20 \times 40 + 20 \times 9$

$+ 8 \times 40 + 8 \times 9$

$= 800 + 180 + 320 + 72 = 1{,}372$

3.

a.		b.	
	87		24
	$\times 15$		$\times 71$
$5 \times 7 \rightarrow$	35	$1 \times 4 \rightarrow$	4
$5 \times 80 \rightarrow$	400	$1 \times 20 \rightarrow$	20
$10 \times 7 \rightarrow$	70	$70 \times 4 \rightarrow$	280
$10 \times 80 \rightarrow$	$+800$	$70 \times 20 \rightarrow$	$+1400$
	1305		1704
c.		d.	
	38		52
	$\times 92$		$\times 65$
$2 \times 8 \rightarrow$	16	$5 \times 2 \rightarrow$	10
$2 \times 30 \rightarrow$	60	$5 \times 50 \rightarrow$	250
$90 \times 8 \rightarrow$	720	$60 \times 2 \rightarrow$	120
$90 \times 30 \rightarrow$	$+2700$	$60 \times 50 \rightarrow$	$+3000$
	3496		3380

4.

a. 55	b. 81	c. 73	d. 99
$\times 12$	$\times 64$	$\times 80$	$\times 11$
10	4	0	9
100	320	0	90
50	60	240	90
$+500$	$+4800$	$+5600$	$+900$
660	5184	5840	1089

5.

a.
$$
\begin{array}{r}
24 \\
\times 17 \\
\hline
28 \\
140 \\
40 \\
+200 \\
\hline
408
\end{array}
$$

b.
$$
\begin{array}{r}
44 \\
\times 39 \\
\hline
36 \\
360 \\
120 \\
+1200 \\
\hline
1716
\end{array}
$$

Multiplying Two-Digit Numbers in Parts, continued

6.

a.
```
    6 2
  × 3 3
  ─────
      6
  1 8 0
    6 0
+ 1 8 0 0
  ─────
  2 0 4 6
```

b.
```
    4 7
  × 5 3
  ─────
    2 1
  1 2 0
  3 5 0
+ 2 0 0 0
  ─────
  2 4 9 1
```

c.
```
    8 3
  × 2 9
  ─────
    2 7
  7 2 0
    6 0
+ 1 6 0 0
  ─────
  2 4 0 7
```

Multiply by Whole Tens in Columns, p. 135

1. a. 5,220
   b. 2,040 (first multiply 51×4)
   c. 1,980 (First multiply 66×3)

2. a. 20×65 kg = 1,300 kg
   b. 4×25 = 100 apples in each crate
   c. He got three crates, which is 300 apples because a crate weighs 20 kg and there are 100 apples per crate.

3. a. 36,800   b. 25,500   c. 4,360
   d. 8,400   e. 134,000   f. 64,800

4. In his five-day work week he drives 5×250 km = 1,250 km. In a month he drives $4 \times 1,250$ km = 5,000 km.

5. He jogged a total of 7×800 m = 5,600 m or 5 km 600 m.

6. 43,500

Puzzle corner. The answer to $14 \times 16 \times 45 \times 50$ is $2 \times 2 \times 5 \times 5 = 100$ times the answer to $7 \times 8 \times 9 \times 10$.

Therefore, the answer is $100 \times 5,040 = $ 504,000.

Multiplying in Parts: Another Way, p. 137

1. b. 40×73 and 8×73 c. 10×42 and 9×42 d. 50×89 and 5×89

2. a. $20 \times 16 = 320$; $8 \times 16 = 128$; $320 + 128 = 448$ b. $40 \times 73 = 2{,}920$; $8 \times 73 = 584$; $2{,}920 + 584 = 3{,}504$
   c. $10 \times 42 = 420$; $9 \times 42 = 378$; $420 + 378 = 798$ d. $50 \times 89 = 4{,}450$; $5 \times 89 = 445$; $4{,}450 + 445 = 4{,}895$

3. a. 40×41 and 6×41
      $40 \times 41 = 1{,}640$; $6 \times 41 = 246$; $1{,}640 + 246 = 1{,}886$

   b. 20×39 and 8×39
      $20 \times 39 = 780$; $8 \times 39 = 312$; $780 + 312 = 1{,}092$

   c. $10 \times 27 + 5 \times 27$
      $10 \times 27 = 270$; $5 \times 27 = 135$; $270 + 135 = 405$

   d. $90 \times 16 + 3 \times 16$
      $90 \times 16 = 1{,}440$; $3 \times 16 = 48$; $1{,}440 + 48 = 1{,}488$

4. $12 \times \$27 = \$324$. To multiply in parts, multiply $10 \times \$27 = \$270$, then $2 \times \$27 = \$54$, and add those.
   Or, solve it this way: for two months, the bill is \$54. For four months, it is \$108. Then multiply that times
   3 to get \$324. Estimates vary. For example, $12 \times \$30 = \$360$.

5. 720 Estimates vary. For example, write $1 \times 2 \times 3 \times 4 \times 5 \times 6$ as 24×30.
   Estimate that as $20 \times 30 = 600$, or as $25 \times 30 = 750$.

6. 12×15 kg $= 180$ kg. To multiply in parts, multiply either 10×15 and 2×15, OR 10×12 and 5×12.
   Estimates vary. For example, 10×15 kg $= 150$ kg.

7. $12 \times \$35 = \$420$. To multiply in parts, multiply either 10×35 and 2×35, OR 30×12 and 5×12.
   Estimates vary. For example, $10 \times \$35 = \$350$, or $12 \times \$40 = \$480$.

The Standard Multiplication Algorithm with a Two-Digit Multiplier, p. 139

1. a. $8 \times 65 = 520$. $520 + 650 = 1{,}170$ b. $4 \times 82 = 328$. $328 + 7{,}380 = 7{,}708$
   c. $20 \times 93 = 1{,}860$. $1{,}860 + 186 = 2{,}046$ d. $50 \times 70 = 3{,}500$. $3{,}500 + 210 = 3{,}710$

2. a. 636 b. 385 c. 6,396 d. 980 e. 494 f. 950 g. 884 h. 931

3. a. Estimate: $60 \times 10 = 600$. Answer 728
   b. Estimate: $30 \times 60 = 1{,}800$ OR $20 \times 70 = 1{,}400$. Answer: 1,625. Here, it helps to round one factor down and one
      up, since both numbers end in 5. (If you round both up, you get $30 \times 70 = 2{,}100$, which is off a lot from the answer).
   c. Estimate: $70 \times 40 = 2{,}800$. Answer: 2,860
   d. Estimate: $45 \times 10 = 450$. Answer: 616
   e. Estimate: $90 \times 50 = 4{,}500$. Answer: 4,232
   f. Estimate: $90 \times 80 = 7{,}200$. Answer: 7,161

4. a. Estimate: $80 \times 80 = 6{,}400$. Answer: 6,561
   b. Estimate: $100 \times 30 = 3{,}000$. Answer: 3,201
   c. Estimate: $30 \times 50 = 1{,}500$. Answer: 1,512

5. a. $15 \times 12 = 180$ eggs. Estimate: $15 \times 10 = 150$
   b. $21 \times 60 = 1{,}260$ minutes. Estimate: $20 \times 60 = 1{,}200$
   c. $11 \times 39 = 429$. No, 11 buses are not enough. Estimate: $11 \times 40 = 440$.
   d. $12 \times 21 = \$252$. Estimate: $12 \times \$20 = \$240$.

6. a. No, it is not. $53 \times 61 = 3{,}233$, and $51 \times 63 = 3{,}213$. They are close though -- the difference is only 20!

   b. No, the answers are not the same. $42 \times 71 = 2{,}982$, and $41 \times 72 = 2{,}952$. The difference is 30.

7. The change is $\$300 - 15 \times \$17 = \underline{\$45}$.

8. She will pay $52 \times \$98 = \underline{\$5{,}096}$.

9. a. 3,600; 0   b. 300; 12,100

Puzzle Corner:

$3 \times 1\boxed{0}5 = 315$; $\boxed{6} \times 6\boxed{6}7 = 4002$; $4 \times 2\boxed{3}4 = 9\boxed{3}6$; $5 \times 8\boxed{3}4 = 4{,}1\boxed{7}0$

Mixed Review Chapter 3, p. 143

1. a. 1,400 b. 4,000 c. 5,200
   d. 340 e. 200 f. 9,000

2.

	$\xleftarrow{\hspace{1.2cm}}$ 126 $\xrightarrow{\hspace{1.2cm}}$	
a.	57	69

$x = 126 - 57 = 69$

	$\xleftarrow{\hspace{1.2cm}}$ 2000 $\xrightarrow{\hspace{1.2cm}}$	
b.	1199	801

$x = 2,000 - 1,199 = 801$

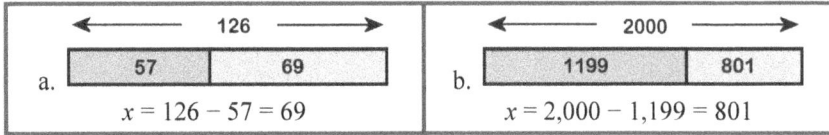

3. a. 440,000 b. 220,000 c. 617,100
   d. 200,000 e. 300,000 f. 81,000

4. a. 7,600; 4,000
   b. 6,190; 26,700
   c. 98,000; 430,000

5. a. 156; 80; 84
   b. 70; 150; 464
   c. 21; 9,990; 1,000

6. a. 119; 980
   b. 0; 700

7. a. < b. = c. <

8. Each time, Mason forgets to add the regrouped tens digit. Correct answers: 336, 384, 717.

9. a. 3 × $345 + $345 = $1,380 total
   b. $145,600 + $12,390 = $157,990. $157,990 × 3 = $473,970 total costs for June, July, and August.
      No, the total cost is not more than half a million dollars.

Review Chapter 3, p. 145

1. a. 1,200; 180
   b. 4,200; 3,300
   c. 81,000; 40,000

2. a. 80; 7 b. 4; 400 c. 300; 80

3. a. 40 b. 2 c. 40

4. In about 8 weeks. 8 × $500 = $4,000.

5. a. Estimation: 7 × 50 = 350; Exact: 336
   b. Estimation: 6 × 800 = 4,800; Exact: 4,878
   c. Estimation: 20 × 20 = 400; Exact: 378
   d. Estimation: 4 × 6,000 = 24,000; Exact: 23,612

6.

Roses	1	2	3	4	5	6	7	8
Price	$0.90	$1.80	$2.70	$3.60	$4.50	$5.40	$6.30	$7.20

7. 2 × 98 = 196; 8 × 17 = 136; 196 − 136 = 60

8. a. 2,000 b. 0 c. 80 d. 20,000

9.

a. 8×24
$= 8 \times 20 + 8 \times 4$
$= 160 + 32 = 192$

```
        20              4
   ┌──────────────┬─────────┐
 8 │   8 x 20     │  8 x 4  │
   └──────────────┴─────────┘
```

b.

```
      3 5
    × 3 9
    ───────
      4 5
    2 7 0
    1 5 0
  + 9 0 0
    ───────
    1 3 6 5
```

```
         30          9
    ┌───────────┬────────┐
    │           │        │
 30 │           │        │
    │           │        │
    ├───────────┼────────┤
  5 │           │        │
    └───────────┴────────┘
```

10. a. He has $50 \times 20 = 1,000$ shirts. The cost is $1,000 \times \$2 = \$2,000$.

Or, write a single number sentence $50 \times 20 \times \$2 = \$2,000$.

b. $8 \times \$2.35 = \$18.80$; $\$20 - \$18.80 = \$1.20$. Or, $\$20 - 8 \times \$2.35 = \$1.20$.

c. $5 \times \$1.50 + \$12.50 = \$20$

d. $\$45 - \$8 = \$37$. $5 \times \$37 = \$185$. Or, $5 \times (\$45 - \$8) = \$185$.

11. a. Two miles.

Minutes	Miles
5	1
10	2
15	3

b. They would weigh 600 g.

Cans	Weight
1	60 g
7	420 g
10	600 g

Chapter 4: Time and Measuring

Time Units, p. 153

1.   a.   b.   c.

Days	Hours
1	24
2	48
3	72
4	96
5	120
6	144
7	168
8	192

Minutes	Seconds
1	60
2	120
3	180
4	240
5	300
6	360
7	420
8	480

Years	Months
1	12
2	24
3	36
4	48
5	60
6	72
7	84
8	96

2. a. He has $6 \times \$120 - \$399 = \$321$ left of his savings.   b. You spend $3 \times 365 = \$1,095$.

3. a. 300 min; 600 min; 720 min
   b. 246 min; 217 min; 470 min
   c. 498 min; 1,210 min; 723 min

4. a. 8 hours 45 min   b. 10 hours 55 min   c. 7 hours 32 min   d. 15 hours 37 min

5. a. $7 \times 35 = 245$ min or 4 hours 5 min.
   b. 2 hours 30 min + 3 hours 50 min + 1 hour 10 min + 3 hours 25 min. = 10 hours 55 min.

6. a. 45 min + 35 min + 1 h 10 min + 1 h 5 min + 40 min = 4 hours 15 min.
   b. 2×40 min = 80 min or 1 hour 20 min. 1 hour 20 min + 3 hours = 4 hours 20 min total.
   c. 2 minutes = 120 seconds. Jean's finishing time was $120 - 24 = 96$ seconds or 1 min 36 seconds.
   d. 1 day = 24 hours. $3 \times 24 = 72$ hours.
   e. $7 \times 25 = 175$ minutes or 2 hours 55 min. In four weeks' time, you would walk your dog about
      $4 \times 175 = 700$ min or 11 hours 40 min.
   f. There are 60 seconds in one minute and 60 minutes in one hour. Multiply $60 \times 60 = 3,600$ seconds.

Elapsed Time 1, p. 156

1. a. 45 minutes   b. 35 minutes   c. 41 minutes   d. 49 minutes
   e. 38 minutes   f. 39 minutes   g. 39 minutes   h. 49 minutes

2. a. 8 hours   b. 4 hours   c. 2 1/2 hours   d. 6 hours   e. 5 1/2 hours
   f. 4 hours   g. 12 hours   h. 13 hours   i. 10 hours

3. a. From 1:40 to 2:30

From 1:40 till 2:00	20 min
From 2 till 2:30	30 min
Total	**50 min**

   b. From 7:30 AM to 3:10 PM

From 7:30 to 8:00	30 min
From 8:00 to 12:00	4 hr
From 12:00 to 3:00	3 hr
From 3:00 to 3:10	10 min
Total	**7 hr 40 min**

Elapsed Time 1, cont.

4. a. 5 hours and 40 minutes   b. 9 hours and 35 minutes

5. The flight was 3 hours and 35 minutes long.

6. a. Channel 1: 1 hour, Channel 2: 1 hour and 5 minutes, Channel 3: 1 hour and 5 minutes.
   b. Kids TV is 15 minutes longer.
   c. Channel 1 has the longest news, Channel 3 has the shortest news,
      and the difference in minutes is 10 minutes.
   d. Channel 1, Megan watched part of Nature film: Whales,
      Channel 3, she watched all of Nature film: Bees and Honey, and on
      Channel 2, she watched all of Nature film: The Antarctic.

The 24-Hour Clock, p. 159

1. a. 5:40   b. 20:00   c. 18:15   d. 11:04   e. 12:30   f. 16:35   g. 23:55   h. 19:05

2. a. 3:00 p.m.   b. 5:29 p.m.   c. 4:23 a.m.   d. 11:55 p.m.
   e. 2:30 p.m.   f. 10:45 a.m.   g. 4:00 p.m.   h. 9:15 p.m.

3. a. Bus 2   b. Bus 6   c. 1 h 20 min   d. 18:01, or 6:01 PM.   e. 33 minutes   f. 52 minutes
   g. Mark has to arrive at Newmarket at 6:15 at the latest, so Bus 8 that leaves at 17:25 will work.

Elapsed Time 2, p. 161

1. a. 1 hr 30 min   b. 1 hr 40 min   c. 6 hr 41 min   d. 4 hr 40 min   e. 4 hr 2 min   f. 7 hr 30 min

2. a. 3 hr 33 min   b. 2 hr 26 min   c. 5 hr 29 min

3. a. 6 hr 26 min   b. 6 hr 27 min   c. 3 hr 46 min

4. a. 5 hr 45 min   b. 8 hr 16 min
   c. 9 hr 38 min   d. 10 hr 53 min
   e. 7 hr 53 min   f. 6 hr 22 min

5. a. 5 hr 18 min   b. 10 hr 40 min   c. 13 hr 40 min

6.

	Time			Time
Patient 1	8:00 - 8:30		Patient 7	11:40 - 12:10
Patient 2	8:30 - 9:00		Patient 8	12:10 - 12:40
Patient 3	9:00 - 9:30		Patient 9	12:40 - 13:10
break	9:30 - 9:50		break	13:10 - 13:30
Patient 4	9:50 - 10:20		Patient 10	13:30 - 14:00
Patient 5	10:20 - 10:50		Patient 11	14:00 - 14:30
Patient 6	10:50 - 11:20		Patient 12	14:30 - 15:00
break	11:20 - 11:40			

7.

Class	Time		Class	Time
Social Studies	8:00 - 8:50		Lunch	11:35 - 12:15
Math	8:55 - 9:45		History	12:15 - 1:05
Science	9:50 - 10:40		P.E.	1:10 - 2:00
English	10:45 - 11:35			

8. Answers will vary. Please check the student's work.

Elapsed Time 3, p. 164

1. a. 6:10   b. 15:25   c. 18:30   d. 11:15   e. 22:03   f. 6:22
   g. Shift 1: 8 hours 30 min; Shift 2: 8 hours; Shift 3: 9 hours. Each shift overlaps the next by 30 minutes.

2. a. 1:20 p.m.   b. 7:42 p.m.   c. 2:45 p.m.
   d. 3:24 p.m.   e. 10:40 a.m.   f. 13:50 p.m.

3. a. 6:55 b. 2:28 c. 12:05 d. 5:15 e. 6:15 f. 11:50

4. a. They left home at 7:15.
   b. They should leave the city by 16:45.

   c.

	Mo	Wd	Th	Fr	Sa
Start:	17:15	17:03	17:05	17:45	17:12
End:	18:20	18:05	18:12	18:39	18:15
Running time:	1 h 5 min	1 h 2 min	1 h 7 min	54 min	1 h 3 min

   d. His total running time was 5 hours and 11 minutes.

   e. From 8:30 until 12:00 is 3 hours 30 min. Then, from 12:00 until
      17:15 is 5 hours 15 min. 3 hours 30 min + 5 hours 15 min is
      8 hours 45 min. Now, subtract the total amount of time, he
      had off for breaks: 8 hours 45 min − 60 minutes = 7 hours 45 min
      of actual work time.

   f. From 7:30 a.m. until 9 p.m. is 13 hours 30 min. 7 × 13 hours 30 min = 91 hours
      210 min = 94 hours 30 min a week.

   g. The flight was delayed 15 min, so it will arrive 15 min later than 5:10 p.m., at 5:25.

Measuring Temperature: Celsius, p. 167

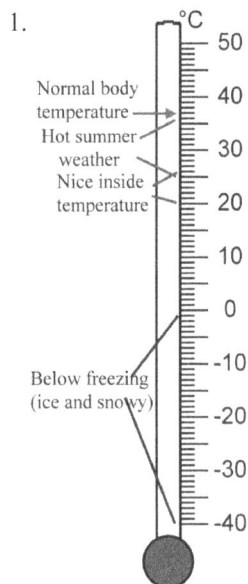

1.

2. a. 20 °C   b. 37°C   c. 12°C   d. 6°C   e. 29°C

3. Answers will vary.

4. Answers will vary.

5.

6.

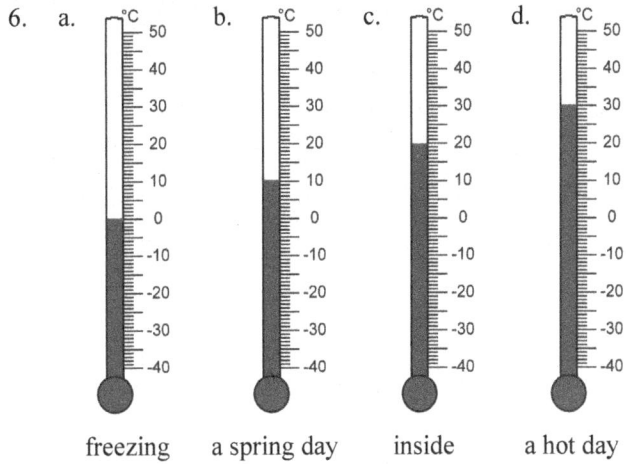

a. freezing   b. a spring day   c. inside   d. a hot day

7. a. -6°C   b. -3°C   c. -11°C   d. -16°C   e. -13°C

8.

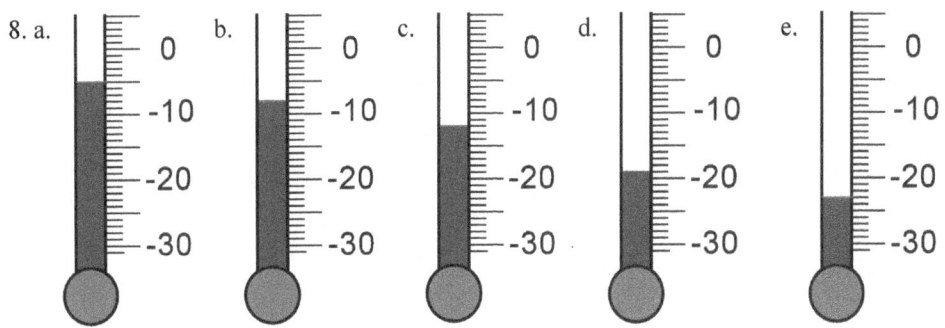

9.

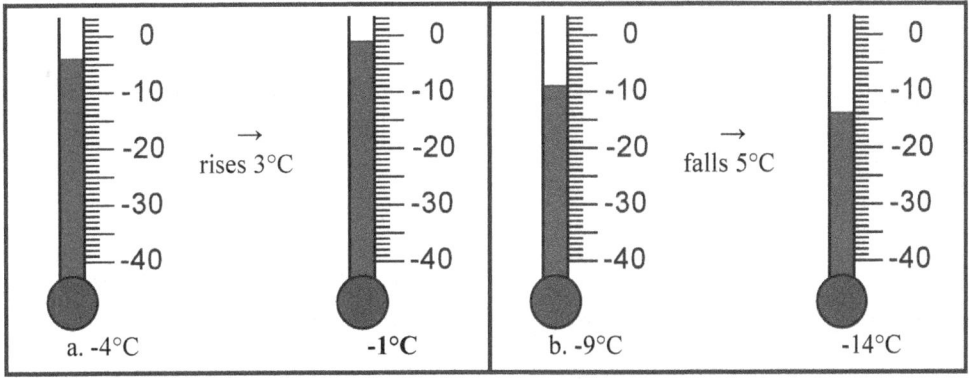

rises 3°C   falls 5°C

a. -4°C   **-1°C**   b. -9°C   -14°C

10. a. -8°C   b. -10°C   c. 2°C   d. -10°C   e. -12°C   f. -2°C   g. 0°C   h. -3°C   i. -16°C

11. a. -15°C   b. The difference is 40°C.

Puzzle corner: 2°C

Measuring Temperature: Fahrenheit, p. 171

1. a. 50°F; Chilly day   b. 81°F; Nice weather   c. 93°F; Very warm day   d. 105°F; Hot desert   e. 73°F; Inside

2. Answers will vary.

3. Answers will vary.

4. Answers will vary. For example:
   a. a very cool autumn morning;   b. a winter day;   c. a very hot summer day.

5. a.   b.   c.   d.   e.

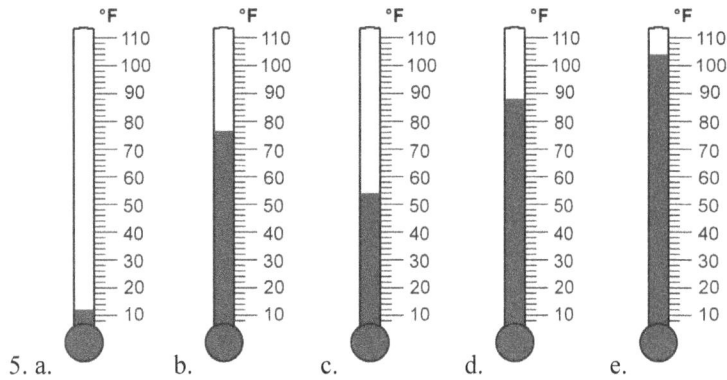

Temperature Line Graphs, p. 173

Month	Jan	Feb	Mar	Apr	May	Jun	Jul	Aug	Sep	Oct	Nov	Dec
Max Temperature	6°C	7°C	10°C	13°C	17°C	20°C	22°C	21°C	19°C	14°C	10°C	7°C

1. a. July
   b. January
   c. March and November; or February and December.
   d. 3 degrees Celsius
   e. 2 degrees Celsius
   f. 16 degrees Celsius

2.

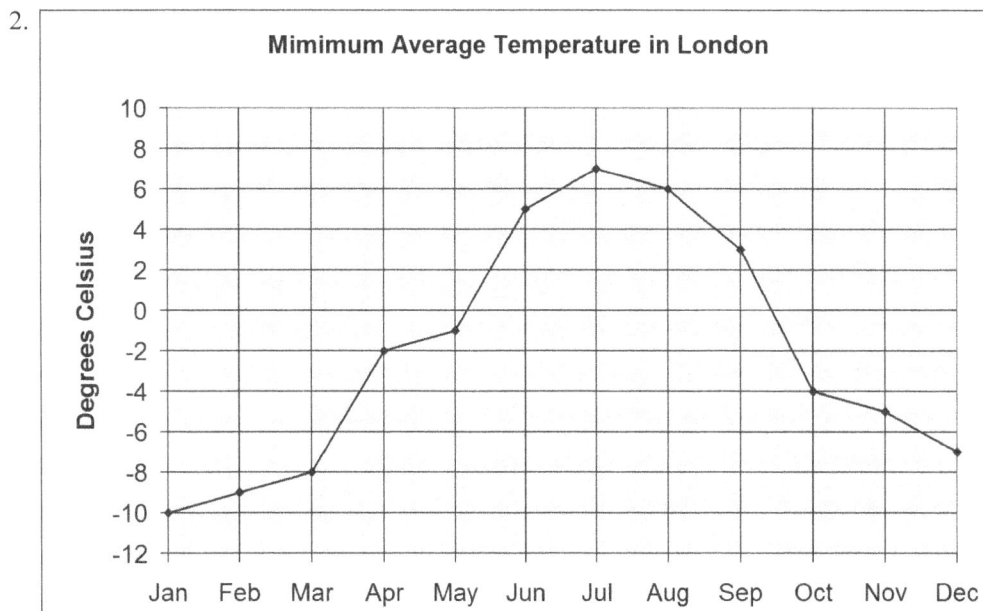

Mimimum Average Temperature in London

a. January   b. 6°   c. 1°   d. 15°

Measuring Length, p. 175

1. a. 1 3/4 in. or 4 cm 5 mm   b. 2 1/2 in. or 6 cm 4 mm   c. 3 1/4 in. or 8 cm 3 mm
   d. 5 1/4 in. or 13 cm 4 mm   e. 4 1/2 in. or 11 cm 5 mm

2. a. 1 1/8 in   b. 2 3/8 in   c. 1 7/8 in   d. 7/8 in   e. 4 3/8 in   f. 3 3/4 in

3. Check students' work.

   a. ————————————

   b. ————————————————

   c. ——————————————————————

   d. exceeds the width of this page

   e. exceeds the width of this page

4. Check students' work.

   a. ————————

   b. ————————————————

   c. ——————

   d. and e. exceed the width of this page.

5. Answers will vary.

More Measuring in Inches and Centimeters, p. 178

1. Answers will vary.

2. Answers will vary.

3. a. 2 1/8 in, 2 1/4 in, 5 in, 3 1/2 in
   b. 5 cm 3 mm, 5 cm 7 mm, 12 cm 8mm, 9 cm

4. a. 3 cm   b. 7 cm   c. 6 inches

5. 1 cm 4 mm

6. a. 10   b. 4   c. 4 1/2 inches   d. 1 1/2 inches   e. 3 inches

7. Answers will vary. Please check the student's work.

Feet, Yards, and Miles, p. 180

1. Answers will vary.

2. Check the student's work.

3. Answers will vary.

4.

a. Feet	a. Inches	b. Feet	b. Inches	c. Feet	c. Inches
1	12	6	72	11	132
2	24	7	84	12	144
3	36	8	96	13	156
4	48	9	108	14	168
5	60	10	120	15	180

Feet, Yards, and Miles, continued

5.

a. 6 ft = <u>72</u> in 11 ft = <u>132</u> in	b. 2 ft 5 in = <u>29</u> in 7 ft 8 in = <u>92</u> in	c. 13 ft 7 in = <u>163</u> in 11 ft 11 in = <u>143</u> in
d. 36 in = <u>3</u> ft 50 in = <u>4</u> ft <u>2</u> in	e. 27 in = <u>2</u> ft <u>3</u> in 100 in = <u>8</u> ft <u>4</u> in	f. 64 in = <u>5</u> ft <u>4</u> in 85 in = <u>7</u> ft <u>1</u> in

6. a. He is 8 inches taller.
   b. She is 5 ft 1 in. tall now.
   c. The difference between their heights is 6 ft 6 in.
   d. The shorter sides measure 2 ft 11 in.

7.

a. Yards	Feet	b. Yards	Feet	c. Yards	Feet
1	3	4	12	7	21
2	6	5	15	8	24
3	9	6	18	9	27

8. a. 18 ft; 39   b. 8 ft; 16 ft   c. 8 yd; 14 yd   d. 4 yd 1 ft; 5 yd 2 ft   e. 7 yd 1 ft; 9 yd 2 ft   f. 10 yd 2 ft; 13 yd 1 ft

9. 400 yard = 1,200 feet, so Jessie ran the longer distance. He ran 1,200 ft − 1,000 ft = 200 feet longer distance.

10. a. ? = 120 ÷ 2 − 18 = 42 yd.   b. 42 × 3 =126 ft

11. 6 yards = 18 ft, so 18 ft − 2 ft − 2 ft = 14 ft of material left, or 4 yd 2 ft.

12. The remaining piece is 1 ft 8 in long.

13 a. 18 ft 4 in
   b. 2 ft 8 in
   c. 8 ft 4 in
   d. 5 ft 6 in + 8 ft 3 in + 3 ft 8 in = 17 ft 5 in, so yes, they will fit.
   e. It is 8 ft tall.

14. a. 21,120 ft   b. 28,750 ft   c. 13,000 ft is longer.   d. About 4 miles.
   e. About 5 1/2 miles.   f. 15,000 feet; about 3 miles.   g. He walks 10 × 950 ft = 9,500 ft; about two miles.

Metric Units for Measuring Length, p. 185

1. and 2. Answers will vary.

3. a. 500 cm; 800 cm; 1,200 cm   b. 406 cm; 919 cm; 1,080 cm   c. 8 m; 2 m 39 cm; 4 m 7 cm

4. a. 50 mm; 80 mm; 140 mm   b. 28 mm; 75 mm; 104 mm   c. 5 cm 0 mm; 7 cm 2 mm; 14 cm 5 mm

5. a. 5,000 m; 23,000 m; 1,200 m   b. 2,800 m; 6,050 m; 13,579 m   c. 2 km; 4 km 300 m; 18 km 700 m

6. a. 14 km 100 m   b. 6 km 400 m   c. 4 km 100 m   d. 4 km 200 m

7. a. 5 m 60 cm   b. 4 cm 4 mm   c. 22 cm 4 mm   d. 1 cm 5mm

8. a. 1,000 mm
   b. 3 km 600 m in a day; 18 km in a week.
   c. Jared is 8 cm taller.

Puzzle corner. She can have 12 complete butterflies on a wall that is 1 meter long,
and 37 complete butterflies on a wall that is 3 meters long.

Customary Units of Weight, p. 188

1. Answers will vary.

2. a. 1 oz   b. 1 lb or 20 oz   c. 4 T   d. 2 T or 3500 lb
   e. 5 oz   f. 130 lb   g. 3 T   h. 22 lb   i. 200 lb

3.

Pounds	1/2	1	2	2 1/2	3	4	5
Ounces	8	16	32	40	48	64	80

Tons	2	3	4	5	10	12	20
Pounds	4,000	6,000	8,000	10,000	20,000	24,000	40,000

4. a. 32 oz; 48 oz; 64 oz   b. 17 oz; 85 oz; 59 oz   c. 68 oz; 45 oz; 83 oz

5. The second baby was 3 oz heavier.

6. 4 apples in 1 lb; 20 apples in 5 lbs.

7. 2 lbs of rice cost $2.88.

8. She put 13 oz of cocoa powder in the bag.

9. a. 1 lb 1 oz; 1 lb 3 oz; 1 lb 7 oz   b. 2 lb; 2 lb 3 oz; 2 lb 14 oz   c. 3 lb 3 oz; 3 lb 7 oz; 3 lb 12 oz

10. a. 8 oz or 1/2 lb   b. 12 oz or 3/4 pound   c. 2 lb 8 oz or 2 1/2 lb
    d. 4 oz or 1/4 lb   e. 5 lb 12 oz or 5 3/4 lb   f. 4 lb 4 oz or 4 1/4 lbs

11. a. 8 oz; 24 oz   b. 4 oz; 36 oz   c. 12 oz; 28 oz

12. a. 1 lb 15 oz   b. 6 lb 11 oz   c. 11 lb 1 oz   d. 8 lb 4 oz

13. a. 3 lb 2 oz   b. 1/4 lb of bread is 4 oz, so Jessie ate more. He ate 3 oz more.
    c. 3 lb 5 oz   d. 4 bags weigh 5 lb; 10 bags weigh 12 lb 8 oz

Metric Units of Weight, p. 192

1. a. 30 kg   b. 2 kg   c. 100 g   d. 20 kg   e. 5 g   f. 50 kg

2.

kilograms	1/2	2	3	3 1/2	5	10	12
grams	500	2,000	3,000	3,500	5,000	10,000	12,000

kilograms	1/2	1	4	4 1/2	6	10	40
grams	500	1,000	4,000	4,500	6,000	10,000	40,000

3. a. 2,000 g; 3,000 g; 4,000 g   b. 1,600 g; 8,080 g; 2,450 g   c. 8,600 g; 5,008 g; 7,041

4. a. 6 kg; 6 kg 700 g; 5 kg 300 g   b. 1 kg 200 g; 6 kg 70 g; 4 kg 770 g

5. a. 3 kg 300 g   b. 6 kg 400 g   c. 10 kg

6. About 7 apples.

7. 6 workbooks.

8. 5 × 400 g is 2000 g , which is 2 kg, and 1 kg + 1 kg is 2 kg, so both
   quantities of chocolate weigh the same.

9. a. 7 kg 520 g   b. 12 kg   c. 7 kg 810 g

10. a. 5 kg 900 g; 2 kg; 9 kg 200 g   b. 1 kg 200 g; 2 kg; 1 kg 100 g

11. 7 kg 350 g total weight.

Metric Units of Weight, cont.

12. 5 kg 800 in total.

13. You need 950 grams more.

14. You need 5 bags. Total cost: $8.45.

Customary Units of Volume, p. 195

1. a. - e. Check the student's answers.

f.

Cups	Ounces
1/2	4
1	8
1 1/2	12
2	16
3	24

2. a.　　　　　　b.　　　　　　c.

Quarts	Cups
1/2	2
1	4
2	8
3	12
4	16
5	20

Quarts	Pints
1/4	1/2
1/2	1
1	2
2	4
3	6
4	8

Gallons	Quarts
1/2	2
1	4
2	8
3	12
4	16
5	20

3. a. 1 quart　b. 4 cups　c. 2 cups　d. 3 pints　e. 1 pint
   f. 2 pints　g. 1 quart　h. 4 cups　i. 1 gallon　j. 2 quarts

4. a. 2 C; 1 C　b. 4 C; 1 C　c. 4 qt; 12 qt　d. 4 pt; 12 C

5. a. 16 oz
   b. The 20-ounce bottle.
   c. The bucket held 3 C.
   d. 2 servings; 6 servings; 4 servings.
   e. A 32-ounce jumbo drink is more.
   f. It is 80 quarts. You need to fill and empty the bucket 8 times.
   g. 20 times
   h. 1 quart is left.

1. a. b. c. d.

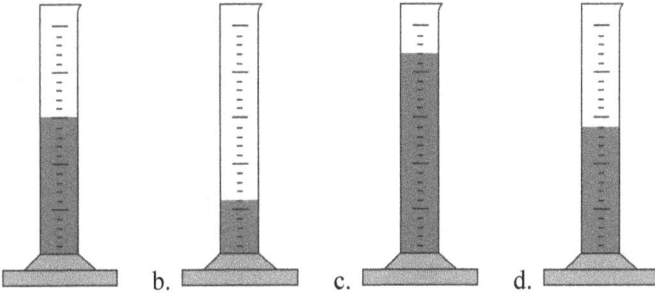

2. a. 5 ml   b. 750 ml   c. 10 L   d. 1 L   e. 200 ml   f. 80 L

3.

L	1/2	1	1 1/2	2	5	12
ml	500	1,000	1,500	2,000	5,000	12,000

L	2 1/2	3	4 1/2	8	10	20
ml	2,500	3,000	4,500	8,000	10,000	20,000

4. a. 2,000 ml; 6,000 ml
   b. 1,200 ml; 4,230 ml
   c. 7,070 ml; 4,330 ml
   d. 3 L; 10 L
   e. 4 L 300 ml; 9 L 880 ml
   f. 3 L 40 ml; 5 L 53 ml

5. a. 750 ml
   b. 1 L 200 ml
   c. 5 glasses; 25 glasses
   d. $4.80

6. a. 5 L 150 ml   b. 7 L 700 ml   c. 3 L 50 ml

7. a. 4 L 400 ml; 9 L 200 ml
   b. 4 L 400 ml; 1 L 700 ml

8. 1 1/2 L + 400 ml + 200 ml = 1,500 ml + 400 ml + 200 ml = 2,100 ml = 2 L 100 ml

9. 5×250 ml $+ 2 \times 2$ L $+ 3 \times 350$ ml $ = $ 1,250 ml $+ 4$ L $+ 1,050$ ml $= 6,300$ ml $ = $ 6 L 300 ml

10. You will buy 7 containers, which will cost $5.46 in total.

Mixed Review Chapter 4, p. 201

1. $24 \times 36 = 20 \times 30 + 20 \times 6 + 4 \times 30 + 4 \times 6 = 600 + 120 + 120 + 24 = 864$ sq units.

2. a. Estimation: $60 \times 30 = 1,800$; Answer: 1,798.
   b. Estimation: $400 \times 8 = 3,200$; Answer: 3,320.
   c. Estimation: $57 \times 100 = 5,700$; Answer: 5,643.
   d. Estimation: $7 \times 700 = 4,900$; Answer: 4,669.

3. a. $3,120   b. 240 km

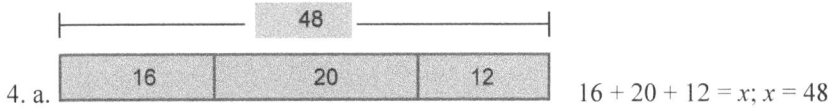

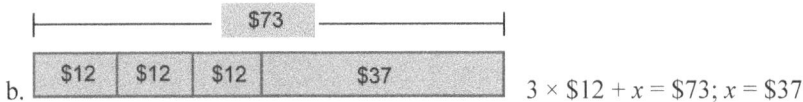

4. a.

48

16	20	12

$16 + 20 + 12 = x; x = 48$

b.

$73

$12	$12	$12	$37

$3 \times \$12 + x = \$73; x = \$37$

5. a. They sold the most strawberries during week 26. About $4,500.
   b. They sold the least strawberries during week 23. The sales were $1,500.
   c. About $12,000.

Review Chapter 4, p. 203

1. a. 6 hours and 52 minutes
   b. 10 hours and 40 minutes

2. The plane lands at 6:10 p.m.

3. a. A cold winter day.   b. A nice temperature for indoors.

4. Check the student's work.

   a. 2 3/8 in. ▬▬▬▬▬▬▬▬▬▬▬▬▬▬▬

   b. 36 mm ▬▬▬▬▬▬▬▬▬

5. a. 150 mm; 68 mm   b. 12 ft; 68 in   c. 425 cm; 8,000 m

6. 16 ft 10 in

7. 5 km 600 m total per week

8. a. 16 kg or 34 lb   b. 2 kg   c. 2 oz

9. a. 112 oz, 91 oz
   b. 6,200 lb, 7,500 g
   c. 2,500 g, 3 kg 456 g

10. He weighed 20 kg 850 g before.

11. The cat food will last five days with 2 ounces left over.

12. a. 3 gal   b. 3 cups   c. 1/2 gal

13. a. 2,300 ml, 6 L 550 ml
    b. 6 pt, 12 cups
    c. 16 qt, 16 fl oz

14. It provided 48 cups of punch. (Three gallons is 12 quarts, and that is 48 cups.)

Puzzle corner: She paid $64.00. A quart has 4×8 oz = 32 oz. The total cost is then 32 oz × $2/oz = $64.

Math Mammoth Grade 4-B
Answer Key

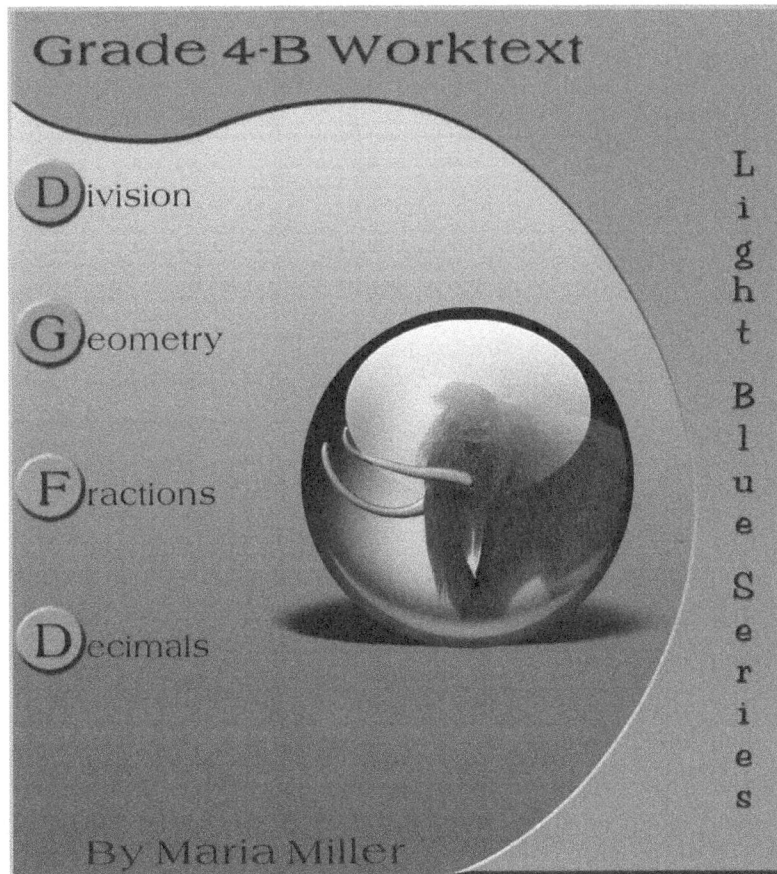

Grade 4-B Worktext

Division

Geometry

Fractions

Decimals

Light Blue Series

By Maria Miller

By Maria Miller

Math Mammoth Grade 4-B Answer Key

Contents

	Work-text page	Answer key page

Chapter 5: Division

	Work-text page	Answer key page
Review of Division	10	60
Division Terms and Division with Zero	13	60
Dividing with Whole Tens and Hundreds	15	61
Order of Operations and Division	18	62
The Remainder, Part 1	20	62
The Remainder, Part 2	23	63
The Remainder, Part 3	25	64
Long Division 1	27	65
Long Division 2	31	65
Long Division 3	34	65
Long Division with 4-Digit Numbers	38	65
More Long Division	42	66
Remainder Problems	45	66
Long Division with Money	49	66
Long Division Crossword Puzzle	51	67
Average	52	67
Finding Fractional Parts with Division	55	68
Problems with Fractional Parts	58	69
Problems to Solve	60	69
Divisibility	63	69
Prime Numbers	67	71
Finding Factors	70	72
Mixed Review Chapter 5	72	73
Review Chapter 5	74	73

Chapter 6: Geometry

	Work-text page	Answer key page
Review: Area of Rectangles	81	75
Problem Solving: Area of Rectangles	84	75
Review: Area and Perimeter	86	76
Lines, Rays, and Angles	90	77
Measuring Angles	93	77
Drawing Angles	100	78
Estimating Angles	102	78

Geometry, continued

	Work-text page	Answer key page
Angle Problems	107	79
Parallel and Perpendicular Lines	112	80
Parallelograms	117	81
Triangles	120	82
Line Symmetry	124	82
Mixed Review Chapter 6	127	83
Review Chapter 6	129	84

Chapter 7: Fractions

	Work-text page	Answer key page
One Whole and Its Fractional Parts	137	85
Mixed Numbers	140	86
Mixed Numbers and Fractions	144	87
Adding Fractions	147	87
Adding Mixed Numbers	149	88
Equivalent Fractions	152	89
Subtracting Fractions and Mixed Numbers	157	92
Comparing Fractions	161	93
Multiplying Fractions by Whole Numbers	165	94
Practicing with Fractions	168	95
Mixed Review Chapter 7	170	96
Review Chapter 7	172	96

Chapter 8: Decimals

	Work-text page	Answer key page
Decimal Numbers—Tenths	177	97
Adding and Subtracting with Tenths	179	97
Two Decimal Digits—Hundredths	181	98
Add and Subtract Decimals in Columns	185	99
Add and Subtract Decimals Mentally	188	100
Using Decimals with Measuring Units	192	101
Mixed Review Chapter 8	194	102
Review Chapter 8	196	103

Chapter 5: Division

Review of Division, p. 10

1. a. $3 \times 4 = 12$; $12 \div 3 = 4$; $12 \div 4 = 3$
   b. $5 \times 3 = 15$; $15 \div 5 = 3$; $15 \div 3 = 5$
   c. $2 \times 4 = 8$; $8 \div 2 = 4$; $8 \div 4 = 2$

2. a. $21 \div 7 = 3$; $21 \div 3 = 7$; $7 \times 3 = 21$; $3 \times 7 = 21$
   b. $24 \div 4 = 6$; $24 \div 6 = 4$; $4 \times 6 = 24$; $6 \times 4 = 24$
   c. $36 \div 4 = 9$; $36 \div 9 = 4$; $9 \times 4 = 36$; $4 \times 9 = 36$

3. a. 8, 9, 10, $22 \div 2 = 11$, $24 \div 2 = 12$, $26 \div 2 = 13$
   b. 9, 8, 7, $30 \div 5 = 6$, $25 \div 5 = 5$, $20 \div 5 = 4$
   c. 9, 10, 11, $120 \div 10 = 12$, $130 \div 10 = 13$, $140 \div 10 = 14$
   d. 8, 7, 6, $35 \div 7 = 5$, $28 \div 7 = 4$, $21 \div 7 = 3$

4.

Eggs	6	12	24	36	42	54	66	78
Omelets	1	2	4	6	7	9	11	13

Thumbtacks	8	24	32	48	64	80	96	104
Pictures	1	3	4	6	8	10	12	13

5. b. $\$45 - \$34 = \$11$, Jim needs $11 more.
   c. $400 \div 4 = 100$; each box has 100 apples.
   d. $24 \div 6 = 4$; each person got 4 pieces.
   e. $5 \times 50 = 250$ total books.
   f. $2 \times \$13 = \$26$; Mom paid $26 for both books.
   g. $20 \div 4 = 5$; there are 5 cows.
   h. $60 \div 3 = 20$; 20 books are on each shelf.

6. a. 9, 10, 5   b. 6, 6, 8   c. 4, 8, 8   d. 8, 3, 5

7. b. $x = 5$ c. $x = 45$ d. $x = 54$

8. a. $10 \times 3 = N$ OR $N = 10 \times 3$; $N = 30$
   b. $9 \times 4 = x$ OR $x = 9 \times 4$; $x = 36$
   c. $20 \times T = 60$, OR $60 = 20 \times T$; $T = 3$
   d. $9 \times y = 81$ OR $81 = y \times 9$; $Y = 9$

9. a. $21 \div 3 = 7$ OR $7 \times 3 = 21$; you can buy 7 books.
   b. $100 \div 5 = 20$ OR $20 \times 5 = 100$; there were
      20 apples in each box.
   c. $\$30 \div 5 = \$6$ OR $5 \times \$6 = 30$; each box costs $6.
   d. $8 \times 5 = 40$; the chocolate bar has 40 squares.
   e. $45 \div 5 = 9$ OR $9 \times 5 = 45$; there are nine fives in 45.
   f. $5 \times 12 = 60$; the boxes weigh 60 pounds.

Division Terms and Division with Zero, p. 13

1. a. 2, the divisor is missing.
   b. 35, the dividend is missing.
   c. 12, the quotient is missing.

2. a. $x \div 7 = 3$; $x = 21$
   b. $140 \div y = 7$; $y = 20$
   c. $150 \div 5 = z$; $z = 30$

3. Answers will vary:
   a. $24 \div 4 = 6$, $30 \div 5 = 6$, $60 \div 10 = 6$
   b. $24 \div 2 = 12$, $24 \div 3 = 8$, $24 \div 6 = 4$

4.

Numbers	Product (written)	Product (solved)	Quotient (written)	Quotient (solved)
12 and 3	12×3	36	$12 \div 3$	4
10 and 5	10×5	50	$10 \div 5$	2
20 and 4	20×4	80	$20 \div 4$	5
100 and 10	100×10	1,000	$100 \div 10$	10

5. a. 8, 0, 1
   b. 11, xx, 1
   c. 50, 0, xx
   d. 0, 1, xx

6. a. $x = 64$
   b. $T = 1$
   c. there are many solutions.
      In fact, x can be any number except 0.
   d. $y = 18$

7. Answers will vary. Examples:
   a. $24 \div 24 = 1$, $4 \div 4 = 1$
   b. $0 \div 36 = 0$, $0 \div 12 = 0$

Puzzle corner. The dividend and quotient both were zeros.
For example, he could have had the problems $0 \div 6 = 0$
and $0 \div 9 = 0$.

1.

a. $300 \times 7 = 2,100$	b. $50 \times 800 = 40,000$	c. $60 \times 40 = 2400$
$2100 \div 7 = 300$	$40000 \div 50 = 800$	$2400 \div 60 = 40$
$2100 \div 300 = 7$	$40000 \div 800 = 50$	$2400 \div 40 = 60$

2. a. 50, 5, 5, 50
   b. 1,000, 100, 10, 10
   c. 6, 60, 6, 60

3. a. 90, 90
   b. 900, 900
   c. 70, 70

4. a. 40, 4, 400
   b. 9, 90, 90
   c. 60, 60, 6000

5. a. 213   b. 4,022   c. 3,101
   d. 110   e. 1,002   f. 1,410

Finding half...	...is the same as dividing by 2!
$\frac{1}{2}$ of 280 is <u>140</u>	$280 \div 2 = \underline{140}$

6. a. 40   b. 12,000   c. 330   d. 2,100

7. Dad's paycheck was: $806 + $806 = $1,612.

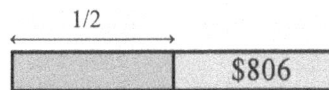

8. The fisherman had: $1/2 \times 800$ kg $- 350$ kg $= 50$ kg left.

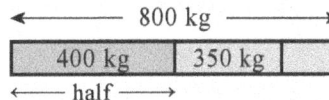

9. She had $2 \times (\$12 + \$15) = \underline{\$54}$ in the beginning.

10.

a. $352 \div 5$	b. $198 \div 4$	c. $403 \div 8$
$\approx 350 \div 5 = 70$	$\approx 200 \div 4 = 50$	$\approx 400 \div 8 = 50$

11.

a. $802 \div 21$	b. $356 \div 61$	c. $596 \div 32$
$\approx 800 \div 20 = 40$	$\approx 360 \div 60 = 6$	$\approx 600 \div 30 = 20$

12.

a. $\approx 80 \div 20 = 4$	b. $\approx 45 \div 5 = 9$
$\approx 120 \div 60 = 2$	$\approx 16,000 \div 400 = 40$
$\approx 2,000 \div 500 = 4$	$\approx 300 \div 30 = 10$

13. $450 \div 5 = 90$

14. Answers will vary but should have a divisor of zero. For example: $67 \div 0$.

15. a. $y = 8,000$ b. $s = 4,200$ c. $w = 30$

16.

a. $500 \div 5 = 100$	b. $466 \div 2 = 233$	c. $366 \div 3 = 122$
$505 \div 5 = 101$	$468 \div 2 = 234$	$369 \div 3 = 123$
$510 \div 5 = 102$	$470 \div 2 = 235$	$372 \div 3 = 124$
$515 \div 5 = 103$	$472 \div 2 = 236$	$375 \div 3 = 125$
$520 \div 5 = 104$	$474 \div 2 = 237$	$378 \div 3 = 126$

Order of Operations and Division, p. 18

1. a. 3   b. 100   c. 120   d. 2,000

2. a. 62   b. 152   c. 2,000   d. 18

3. a. 9   b. 17   c. 200   d. 5

4.

a.	b.	c.
$24 \div 2 + 10 = 22$	$18 + 30 \div 2 = 33$	$40 - 40 \div 8 = 35$
$24 \div (2 + 10) = 2$	$(18 + 30) \div 2 = 24$	$(40 - 40) \div 8 = 0$

5. a. $(20 + 15) \div 5 = 7$
   b. $20 - 50 \div 5 = 10$
   c. $20 \times 30 - 100 = 500$

6. $(21 + 17) \div 2$. The answer is 19 figures.

7. $6 \times 6 \div 4$. The answer is $9.

8. a. 5; 7   b. 60; 120   c. 20; 20   d. 1; 1   e. 0; 0

9. a. $5 \div 5 \times 5 = 5$
   b. $(5 - 5) \times 5 = 0$
   c. $(5 + 5) \div 5 = 2$
   d. $(5 + 5) \times (5 + 5) = 100$
   e. $5 \times 5 + 5 - 5 = 25$
   OR $5 \times 5 - 5 + 5 = 25$
   OR $5 - 5 + 5 \times 5 = 25$
   OR $5 \times 5 - (5 - 5) = 25$

Puzzle corner:
$(5 - 5) \times 5 = 0$
$5 \div 5 = 1$
$(5 + 5) \div 5 = 2$
$(5 + 5 + 5) \div 5 = 3$
$(5 \times 5 - 5) \div 5 = 4$
$5 \times 5 \div 5 = 5$
$(5 \times 5 + 5) \div 5 = 6$
$(5 \times 5 + 5 + 5) \div 5 = 7$
$(5 + 5) - (5 + 5) \div 5 = 8$
$(5 + 5) - (5 \div 5) = 9$
$5 + 5 = 10$

The Remainder, Part 1, p. 20

1.

a. Divide into groups of 4.	b. Divide into groups of 2.	c. Divide into groups of 5.
$10 \div 4 = 2$ R2	$17 \div 2 = 8$ R1	$12 \div 5 = 2$ R2

2. a. $14 \div 4 = 3$ R2   b. $7 \div 3 = 2$ R1   c. $19 \div 6 = 3$ R1

3.

a. Divide 16 into groups of 5.	b. Divide 17 into groups of 3.	c. Divide 15 into groups of 4.
$16 \div 5 = 3$ R1	$17 \div 3 = 5$ R2	$15 \div 4 = 3$ R3

4.

a. $17 \div 4 = 4$ R1	b. $9 \div 2 = 4$ R1	c. $11 \div 6 = 1$ R5

5. a. $10 \div 3 = 3$ R1   b. $17 \div 5 = 3$ R2   c. $11 \div 4 = 2$ R3

6. a. $13 \div 5 = 2$ R3   b. $18 \div 4 = 4$ R2   c. $10 \div 4 = 2$ R2

7.

a. $27 \div 5 = 5$ R2 5 goes into 27 five times.	b. $16 \div 6 = 2$ R4 6 goes into 16 two times.	c. $11 \div 2 = 5$ R1 2 goes into 11 five times.
d. $37 \div 5 = 7$ R2	e. $26 \div 3 = 8$ R2	f. $56 \div 9 = 6$ R2
g. $43 \div 5 = 8$ R3	h. $34 \div 6 = 5$ R4	i. $40 \div 7 = 5$ R5

8.

a. $23 \div 4 = 5$ R3 $23 \div 5 = 4$ R3	b. $16 \div 7 = 2$ R2 $20 \div 3 = 6$ R2	c. $21 \div 8 = 2$ R5 $12 \div 9 = 1$ R3

9.

a. $10 \div 5 = 2$ R 0 $11 \div 5 = 2$ R1 $12 \div 5 = 2$ R2 $13 \div 5 = 2$ R3 $14 \div 5 = 2$ R4 $15 \div 5 = 3$ R0	b. $17 \div 3 = 5$ R2 $18 \div 3 = 6$ R0 $19 \div 3 = 6$ R1 $20 \div 3 = 6$ R2 $21 \div 3 = 7$ R0 $22 \div 3 = 7$ R1	c. $12 \div 4 = 3$ R0 $13 \div 4 = 3$ R1 $14 \div 4 = 3$ R2 $15 \div 4 = 3$ R3 $16 \div 4 = 4$ R0 $17 \div 4 = 4$ R1

10. a. $27 \div 5 = 5$ R2. He had five rows. Two cars were left over.
   b. $19 \div 5 = 3$ R4. She had five groups of 5. You can make a smaller group with only four children in it.
   c. $(36 - 3) \div 6 = 5$ R3. She has five full bags of cookies.
   d. No, because $51 \div 8 = 6$ R3.
   e. Of four, no. $35 \div 4 = 8$ R3 (the division is not even.) Of five, yes. $35 \div 5 = 7$.
      Of six, no. $35 \div 6 = 5$ R5. Of seven, yes. $35 \div 7 = 5$.
   f. $38 \div 6 = 6$ R2. There were two photos on the last page. Six pages were full.

The Remainder, Part 2, p. 23

1. a. 3   b. 9   c. 7   d. 9

2. a.
$$\begin{array}{r} 6 \\ 5\overline{)3\,2} \\ -3\,0 \\ \hline 2 \end{array}$$
b.
$$\begin{array}{r} 8 \\ 5\overline{)4\,4} \\ -4\,0 \\ \hline 4 \end{array}$$
c.
$$\begin{array}{r} 6 \\ 6\overline{)3\,7} \\ -3\,6 \\ \hline 1 \end{array}$$
d.
$$\begin{array}{r} 4 \\ 7\overline{)2\,9} \\ -2\,8 \\ \hline 1 \end{array}$$

e.
$$\begin{array}{r} 5 \\ 8\overline{)4\,6} \\ -4\,0 \\ \hline 6 \end{array}$$
f.
$$\begin{array}{r} 5 \\ 9\overline{)5\,2} \\ -4\,5 \\ \hline 7 \end{array}$$
g.
$$\begin{array}{r} 8 \\ 4\overline{)3\,5} \\ -3\,2 \\ \hline 3 \end{array}$$
h.
$$\begin{array}{r} 6 \\ 9\overline{)5\,7} \\ -5\,4 \\ \hline 3 \end{array}$$

3. a. $6 \times 5 + 2 = 32$ b. $8 \times 5 + 4 = 44$ c. $6 \times 6 + 1 = 37$ d. $4 \times 7 + 1 = 29$
   e. $5 \times 8 + 6 = 46$ f. $5 \times 9 + 7 = 52$ g. $8 \times 4 + 3 = 35$ h. $6 \times 9 + 3 = 57$

4. $33 \div 6 = 5$ R3. Jill needed six containers, but only five were full.

5. $53 \div 12 = 4$ R5. Mom needed 5 cartons for all the eggs.

6. $36 \div 11 = 3$ R3. She put three pencils back into the cabinet.

7. $58 \div 8 = 7$ R2. They got seven full boxes.

8. $3 \times 23 + 15 = 84$. He had 84 award stickers.

The Remainder, Part 3, p. 25

1. One bus holds 42 children. Two buses hold 84 children. Three buses hold 126 children. So, three buses were needed to hold 100 children

2. She needed four folders. (Three folders is not enough, because $3 \times 20 = 60$. Yet, four *is* enough because $4 \times 20 = 80$.) Three of them were full.

3. a. They could make three classes of 22 first graders. (They don't have enough first graders for four classes since $4 \times 22 = 88$ which is more than 77.)

   b. They will get three classes of 20 first graders, and 17 children in a fourth class.

4. The teams had six, seven, and seven players.

5. a. 7 R5   b. 8 R2   c. 8 R4   d. 4 R3

6.

| a. $12 \div 3 = 4$ R0 | b. $10 \div 2 = 5$ R0 | c. $19 \div 4 = 4$ R3 |

7.

a. $21 \div 5 = 4$ R1	b. $56 \div 8 = 7$ R0	c. $43 \div 7 = 6$ R1
$22 \div 5 = 4$ R2	$57 \div 8 = 7$ R1	$44 \div 7 = 6$ R2
$23 \div 5 = 4$ R3	$58 \div 8 = 7$ R2	$45 \div 7 = 6$ R3
$24 \div 5 = 4$ R4	$59 \div 8 = 7$ R3	$46 \div 7 = 6$ R4

8. The shortcut is: the remainder is always the last digit of the dividend (the number you divide), and the other digits are the quotient (the answer).

a. $29 \div 10 = 2$ R9	b. $78 \div 10 = 7$ R8	c. $54 \div 10 = 5$ R4
$30 \div 10 = 3$ R0	$79 \div 10 = 7$ R9	$55 \div 10 = 5$ R5
$31 \div 10 = 3$ R1	$80 \div 10 = 8$ R0	$56 \div 10 = 5$ R6

Puzzle Corner:
a. $16 \div 5 = 3$ R1 OR $16 \div 3 = 5$ R1
b. $31 \div 7 = 4$ R3 OR $31 \div 4 = 7$ R3
c. $135 \div 30 = 4$ R3 OR $123 \div 4 = 30$ R3

Long Division 1, p. 27

1.

a. Make 2 groups	b. Make 3 groups	c. Make 3 groups	d. Make 4 groups
$\begin{array}{r} 3\ 1 \\ 2\overline{)6\ 2} \end{array}$	$\begin{array}{r} 2\ 1 \\ 3\overline{)6\ 3} \end{array}$	$\begin{array}{r} 1\ 0\ 2 \\ 3\overline{)3\ 0\ 6} \end{array}$	$\begin{array}{r} 1\ 2\ 0 \\ 4\overline{)4\ 8\ 0} \end{array}$

2. a. 21   b. 131   c. 220   d. 2,010   e. 22   f. 3,021   g. 110   h. 1,201

3. a. 41   b. 71   c. 60   d. 31   e. 92   f. 61   g. 611   h. 601   i. 710   j. 901

4.

a. $\begin{array}{r} 3\ 1\ R1 \\ 2\overline{)6\ 3} \end{array}$	b. $\begin{array}{r} 1\ 2\ 2\ R1 \\ 2\overline{)2\ 4\ 5} \end{array}$	c. $\begin{array}{r} 1\ 1\ 2\ R2 \\ 3\overline{)3\ 3\ 8} \end{array}$	d. $\begin{array}{r} 2\ 3\ 0\ R1 \\ 2\overline{)4\ 6\ 1} \end{array}$

5. a. 211 R3   b. 34 R1   c. 122 R1   d. 22 R1   e. 60 R1   f. 300 R5   g. 30 R5   h. 310 R2

6. a. 42 R2   b. 31 R2   c. 711 R1   d. 711 R1   e. 1,101 R2   f. 4,031 R1

7. a. 110, 410   b. 9, 123   c. 412, 6 R20

Long Division 2, p. 31

1. a. 16   b. 24   c. 29   d. 15   e. 19   f. 39   g. 26   h. 47

2. a. 47   b. 43   c. 64   d. 34   e. 84   f. 58

Long Division 3, p. 34

1. a. 115   b. 123   c. 244   d. 276   e. 318   f. 121   2. a. 189   b. 166   c. 142   d. 117   e. 152   f. 117
   g. 113   h. 113   i. 325   j. 113   k. 112   l. 218

Long Division with 4-Digit Numbers, p. 38

1. a. 2,347   b. 2,310   c. 1,785   d. 4,885

2. a. 1,934   b. 551   c. 1,340
   d. 1,138   e. 1,317   f. 1,216

3. a. 493   b. 384   c. 924   d. 49   e. 87   f. 371

4. 9 × $16 ÷ 2 = $72. They each paid $72.

5. 504 min ÷ 7 = 72 min, or 1 h 12 min each day

6. a. 2600 ÷ 8 = 325. The second clue is at 325 feet.
   b. The third clue is at 650 feet.

7. a. 96 ÷ 6 = 16. Sixteen children were coming to the party.
   b. 8 × 25 = 200 and 200 − 96 = 104.
      She had 104 balloons left.

More Long Division, p. 42

1. a. 1,045   b. 1,406   c. 2,037   d. 1,307

2. a. 2,705   b. 1,308   c. 1,309   d. 1,063

3. a. 108   b. 205   c. 402   d. 405   e. 308   f. 1,070

4. a. $285 \div 5 = 57$.
   There are 57 buttons in one compartment.
   b. $3 \times 57 = 171$.
   There are 171 buttons in three compartments.

4. c. $4 \times 57 = 228$.
   There are 228 buttons in four compartments.

5. The payments were ($9,620 − $2,000) ÷ 4 = $1,905 each.

6. a. 21,234   b. 35,407   c. 21,645   d. 3,162   e. 5,275

Remainder Problems, p. 45

1. a. 171 R1   Check: $3 \times 171 + 1 = 514$
   b. 84 R1   Check: $8 \times 84 + 1 = 673$
   c. 317 R3   Check: $6 \times 317 + 3 = 1,905$
   d. 2,051 R1 Check: $4 \times 2,051 + 1 = 8,205$

2. a. wrong; 77 R1   b. right   c. wrong: 451
   d. The remainder is larger than the divisor.

3. a. $112 \div 9 = 12$ R4. We get 12 rows, 9 chairs each row, and 4 chairs will be left over or put in an extra row.
   b. $800 \div 3 = 266$ R2. We get 266 piles, 3 erasers in each pile, and 2 erasers left over.

4. They will get 166 full bags.

5. $20 \times 50 = 1,000$ and $19 \times 50 = 950$. So, 19 buses is enough to transport 940 people.

6. $75 \div 4 = 18$ R3. One 18-day vacation and three 19-day vacations. If the division had been even, all of the vacations would have been 18 days, but now there are three extra days to be added to three of the vacations.

7. $400 − (2 \times 90) − (4 \times 40) = 60$; $60 \div 6 = 10$.
   They will have ten full 6-kg boxes of strawberries.

8. a. Yes. There will be 103 containers. $412 \div 4 = 103$.
   b. No, there will be 82 containers with 2 left over.
      $412 \div 5 = 82$ R2.
   c. No, there will be 68 with four left over.
      $412 \div 6 = 68$ R4.

9. $740 \div 6 = 123$ R2. Paint 123 pieces in four of the colors (any four), and 124 pieces in the two remaining colors.

10. a. 70 R1; 70 R2; 71
    b. 172 R2; 172 R3; 172 R4
    c. 82 R1; 82 R2; 82 R3
    d. 798 R3   798 R4   798 R5
    You can figure out the two other problems after solving one, because the remainder will increase by one as the dividend increases by one.

11. It would be 38 R4. The only difference is that the remainder increases by 1.

12. a. 78 R7; 6 R6; 34
    b. 45 R2; 50 R9; 5 R2
    c. 46 R3; 98 R2; 92 R5
    The ones digit of the dividend will always be the remainder.

Puzzle Corner: The remainder is larger than the divisor.

Long Division with Money, p. 49

1. a. $8.47   b. $3.72

2. a. $28.50   b. $1.14

3. $25.56 + $3.55 + $2.75 = $31.86
   $31.86 ÷ 2 = $15.93. Each girl paid $15.93.

4. ($25.95 + $4.35) ÷ 3 = $10.10. Each person's share was $10.10.

5. $358.60 − $100 = $258.60; $258.60 ÷ 4 = $64.65.
   Each payment was $64.65.

6. $12.96 ÷ 8 = $1.62.
   You will pay $1.62. Your brother will pay $1.62.
   Mom will pay $12.96 − $1.62 − $1.62 = $9.72.

Long Division Crossword Puzzle, p. 51

1. Across:
   a. $3,440 \div 8 = 430$
   b. $574 \div 7 = 82$
   c. $234 \div 9 = 26$
   d. $1,707 \div 3 = 569$
   e. $4,756 \div 2 = 2,378$

   Down:
   a. $1,072 \div 8 = 134$
   b. $6,135 \div 3 = 2,045$
   c. $145 \div 5 = 29$
   d. $2,652 \div 4 = 663$
   e. $1,442 \div 7 = 206$
   f. $3,474 \div 9 = 386$

a. 1					
3		b. 2	b. 8	e. 2	
a. 4	3	0		0	
		4	c. 2	6	
		d. 5	d. 6	9	
			6		f. 3
		e. 2	3	7	8
					6

Average, p. 52

1. $(78 + 87 + 69 + 86) \div 4 = 80$. Judith's average score is 80.

2. $(18 + 22 + 26 + 23 + 16) \div 5 = 21$. The average temperature for the day was 21°C.

3. $414 \div 6 = 69$. Dad averaged 69 km in one hour.

4. $12 \times 55 = 660$. A dozen eggs would weigh 660 grams.

5. $7 \times 76 = 532$. It cost $532 for one week.

6. $(234 + 178 + 250 + 198) \div 4 = 215$. Her weekly average grocery bill was $215.

7. The girls' average time was 15 minutes. The boys' average time was 13 minutes. The boys are faster. The difference is two minutes.

8. a.

Quiz score	Frequency
13-15	1
16-18	1
19-21	2
22-24	4
25-27	0
28-30	2

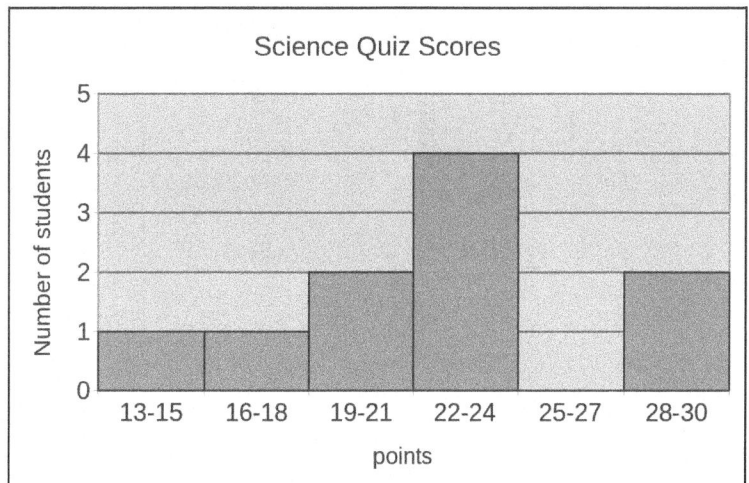
Science Quiz Scores

b. The average score is 22.
c. Look at the "peak" of the graph. The average is usually near that point.

9. a. The average age is 29.   b. Now the average age is 34.

Puzzle corner: $213 \div 12$ is 17 R9.

1.

a.	b.	c.	d.
$10 \div 5 = 2$	$9 \div 3 = 3$	$16 \div 2 = 8$	$15 \div 3 = 5$
$\frac{1}{5}$ of 10 is 2.	$\frac{1}{3}$ of 9 is 3.	$\frac{1}{2}$ of 16 is 8.	$\frac{1}{3}$ of 15 is 5.

2.

a. $30 \div 5 = 6$	b. $48 \div 6 = 8$	c. $25 \div 5 = 5$	d. $50 \div 5 = 10$
$\frac{1}{5}$ of 30 is 6.	$\frac{1}{6}$ of 48 is 8.	$\frac{1}{5}$ of 25 is 5.	$\frac{1}{5}$ of 50 is 10.

3.

a. $\frac{1}{6}$ of 30 is 5.	b. $\frac{1}{7}$ of 49 is 7.	c. $\frac{1}{10}$ of 250 is 25.
$\underline{30} \div \underline{6} = \underline{5}$	$49 \div 7 = 7$	$250 \div 10 = 25$
d. $\frac{1}{2}$ of 480 is 240.	e. $\frac{1}{9}$ of 1,800 is 200.	f. $\frac{1}{5}$ of 400 is 80.
$480 \div 2 = 240$	$1,800 \div 9 = 200$	$400 \div 5 = 80$

4.

a.	b.	c.
$\frac{1}{3}$ of 9 apples is $\underline{3}$ apples.	$\frac{1}{4}$ of 12 flowers is $\underline{3}$.	$\frac{1}{5}$ of $\underline{15}$ fish is $\underline{3}$ fish.
$\frac{2}{3}$ of 9 apples is $\underline{6}$ apples.	$\frac{2}{4}$ of 12 flowers is $\underline{6}$.	$\frac{2}{5}$ of $\underline{15}$ fish is $\underline{6}$ fish.
$\frac{3}{3}$ of 9 apples is $\underline{9}$ apples.	$\frac{3}{4}$ of 12 flowers is $\underline{9}$.	$\frac{3}{5}$ of $\underline{15}$ fish is $\underline{9}$ fish.
	$\frac{4}{4}$ of 12 flowers is $\underline{12}$.	$\frac{4}{5}$ of $\underline{15}$ fish is $\underline{12}$ fish.
		$\frac{5}{5}$ of $\underline{15}$ fish is $\underline{15}$ fish.

5. a. 4, 8, 12   b. 4, 12, 20   c. 50, 150, 350
   d. 70, 140, 350   e. 30, 210, 330   f. 7, 63, 350

6. a. Marsha got $18 from her mom. She put into her savings $6, which was one-<u>third</u> part of it. $18 \div 6 = 3$
   b. Mariana spent one-fourth of her $80 savings, or <u>$20</u>. $80 \div 4 = 20$
   c. One-<u>fifth</u> of all the 25 boys went jogging, which meant that 5 boys went jogging. $25 \div 5 = 5$

7. a. One pound is one-<u>eighth</u> part of the bag. It costs $0.40.
   b. Five-eighths costs $2.00.

8. a. One-tenth of the pie weighs 120 grams.
   b. Nine-tenths of the pie weighs 1,080 grams.

9. $28 \div 4 = 7$, and $7 \times 3 = 21$. Three would cost $21.

Finding Fractional Parts with Division, cont.

10. You can solve this in several ways. One way is to use the idea that ¼ is half of ½ and that ⅛ is half of ¼. The other way is to calculate ½ of $24.40 separately, ¼ of $24.40 separately, and 1/8 of $24.40 separately.

    Mark: $12.20   Judy: $6.10   Art: $3.05   Grace: $3.05

11. Erica and James each had 28 balloons to sell. By the evening, Erica had sold 1/2 or 14 of her balloons. James had sold 3/4 or 21 of his. Together they had sold 35 balloons.

Problems with Fractional Parts, p. 58

1. a. One slice weighs 20 grams.
   b. Three slices weigh 60 grams.
   c. Eleven slices weigh 220 grams.

2. a. Two-sixths of the pie weighs 150 grams.
   b. It weighs 375 grams.

3. If you need to calculate 5/9 of the number 729, first divide 729 by 9, then multiply the result by 5.
   5/9 of 729 is 405

4. $36.50 ÷ 5 × 2 = $14.60

5. The other washer costs $452 ÷ 4 × 3 = $339.

6. 12,600 ÷ 9 × 2 = 2,800
   a. 9,800 miles left   b. 2,800 miles

7. $268 ÷ 4 × 3 = $201. It cost $201. She has $67 left.

8. a. 6 tons, or 12,000 pounds.

9. Edward worked (56 ÷ 4) × 3 = 42 hours.
   James worked 21 hours.
   Together they worked 56 + 42 + 21 = 119 hours.

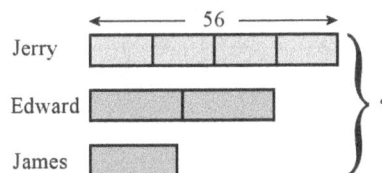

Problems to Solve, p. 60

1. $3.25 + 3 × $3.25 = $13

2. 1,200 are females. Since there are three times as many females as males, we can divide the 1,600 workers into four parts. One-fourth part of 1,600 is 400. So, there are 3 × 400 or 1,200 female workers.

3. Cindy has $14 left. (Half of Cindy's money is $14.)

4. 96 workers. Since 84 is 7/8 of the workers, 84 ÷ 7 = 12 gives you 1/8 of the workers, and then 8 × 12 = 96 is the total amount.

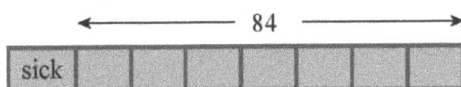

5. Mary got 16 pieces. One-third of the pieces is 8 pieces.

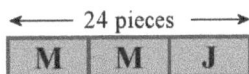

6. 2 yards, 2 feet, 6 inches.
   One-fifth of 15 yards is 3 yards. Now subtract 3 yards − 6 inches = 2 yards 2 feet 6 inches.

7. There are 6 speckled chickens. Solution: there are 18 white chickens (three that were sold and 15 that were left), and that is 3/4 of all the chickens.

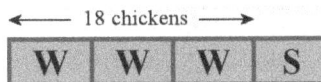

So 18 ÷ 3 = 6 gives you 1/4 of the chickens (one block). And then, one block, or six chickens, are speckled.

8. They would both cost $9 after the discount.
   One-fourth of $12 is $3, so the new price is $9.
   One-third of $13.50 is $4.50, so the new price is $9.

9. Jackie paid $5.00. First find the total without the discount: 5 × $1.50 = $7.50. One-third of that is $2.50. The price with discount is then $7.50 − $2.50 = $5.00

Divisibility, p. 63

1. a. 7; yes   b. 6 R4; no   c. 3 R2; no   d. 12; yes

2. a. 24 R2, no   b. 86 R1; no   c. 418 R2; no

3. Here is a multiplication fact: 8 × 9 = 72. So, 8 is a factor of 72, and so is 9.
   Also, 72 is a multiple of 8, and also 72 is a multiple of 9. And, 72 is divisible by 8 and also by 9.

Divisibility, continued

4.

a. Is 5 a factor of 55? Yes, because $5 \times 11 = 55$.	b. Is 8 a divisor of 45? No, because $45 \div 8 = 5$ R5.
c. Is 36 a multiple of 6? Yes, because $6 \times 6 = 36$.	d. Is 34 a multiple of 7? No, because $34 \div 7 = 4$ R6.
e. Is 7 a factor of 46? No, because $46 \div 7 = 6$ R4. (It is not an even division.)	f. Is 63 a multiple of 9? Yes, because $7 \times 9 = 63$.

5. a. 0, 11, 22, 33, 44, 55, 66, 77, 88, 99, 110, 121, 132, 143, 154

    b. 0, 111, 222, 333, 444, 555, 666, 777, 888, 999, 1,110, 1,221, 1,332, 1,443, 1,554, 1,665

6.

number	divisible by 2	divisible by 5
750	x	x
751		
752	x	
753		
754	x	

number	divisible by 2	divisible by 5
755		x
756	x	
757		
758	x	
759		

number	divisible by 2	divisible by 5
760	x	x
761		
762	x	
763		
764	x	

number	divisible by 2	divisible by 5
765		x
766	x	
767		
768	x	
769		

7.

number	divisible by 2	divisible by 5	divisible by 10
860	x	x	x
861			
862	x		
863			
864	x		

number	divisible by 2	divisible by 5	divisible by 10
865		x	
866	x		
867			
868	x		
869			

number	divisible by 2	divisible by 5	divisible by 10
870	x	x	x
871			
872	x		
873			
874	x		

> If a number is divisible by 10, it ends in zero, so it is ALSO divisible by _2_ and _5_ .

8. a. 2, <u>4</u>, 6, <u>8</u>, 10, <u>12</u>, 14, <u>16</u>, 18, <u>20</u>, 22, <u>24</u>, 26, <u>28</u>, 30, <u>32</u>, 34, <u>36</u>, 38, <u>40</u>, 42, <u>44</u>, 46, <u>48</u>, 50, <u>52</u>, 54, <u>56</u>, 58, <u>60</u>
   This is also a list of multiples of (or multiplication table of) 2.

   b. 2, <u>4</u>, 6, <u>8</u>, 10, <u>12</u>, 14, <u>16</u>, 18, <u>20</u>, 22, <u>24</u>, 26, <u>28</u>, 30, <u>32</u>, 34, <u>36</u>, 38, <u>40</u>, 42, <u>44</u>, 46, <u>48</u>, 50, <u>52</u>, 54, <u>56</u>, 58, <u>60</u>
   These are every other number in the list of multiples of 2.

   c. 2, <u>4</u>, **6**, <u>8</u>, 10, <u>12</u>, 14, <u>16</u>, **18**, <u>20</u>, 22, <u>24</u>, 26, <u>28</u>, **30**, <u>32</u>, 34, <u>36</u>, 38, <u>40</u>, **42**, <u>44</u>, 46, <u>48</u>, 50, <u>52</u>, **54**, <u>56</u>, 58, <u>60</u>
   These are every third number in the list of multiples of 2, or every third even number divisible by 6.

   d. 12, 24, 36, 48, and 60 - or multiples of 12.

9. a. 3, 6, 9, 12, 15, 18, 21, 24, 27, 30, 33, 36, 39, 42, 45, 48, 51, 54, 57, 60
   This is also a list of multiples of (or multiplication table of) 3.

   b. 3, <u>6</u>, 9, <u>12</u>, 15, <u>18</u>, 21, <u>24</u>, 27, <u>30</u>, 33, <u>36</u>, 39, <u>42</u>, 45, <u>48</u>, 51, <u>54</u>, 57, <u>60</u>
   These are every second number in the list of multiples of 3.

   c. 3, <u>6</u>, **9**, <u>12</u>, 15, **<u>18</u>**, 21, <u>24</u>, **27**, <u>30</u>, 33, <u>36</u>, 39, <u>42</u>, **45**, <u>48</u>, 51, <u>54</u>, 57, **<u>60</u>**
   These are every third number in the list of multiples of 3.

10. 18, 36, 54

11. 1

12. It is also a multiple of 1, 2, 10, and 20.

Mystery number: 33 and 60

number	divisible by 1	divisible by 2	divisible by 3	divisible by 4	divisible by 5	divisible by 6	divisible by 7	divisible by 8	divisible by 9	divisible by 10
2	x	x								
3	x		x							
4	x	x		x						
5	x				x					
6	x	x	x			x				
7	x						x			
8	x	x		x				x		
9	x		x						x	
10	x	x			x					x
11	x									
12	x	x	x	x		x				
13	x									
14	x	x					x			
15	x		x		x					
16	x	x		x				x		
17	x									
18	x	x	x			x			x	
19	x									
20	x	x		x	x					x
21	x		x				x			
22	x	x								
23	x									
24	x	x	x	x		x		x		
25	x				x					
26	x	x								
27	x		x						x	
28	x	x		x			x			
29	x									
30	x	x	x		x	x				x
31	x									
32	x	x		x				x		
33	x		x							
34	x	x								
35	x				x		x			

2. Prime numbers: 2, 3, 5, 7, 11, 13, 17, 19, 23, 29, 31

3. Answers will vary, as you can write a composite number as a product in many different ways.

a. 33 is composite. $33 = 3 \times 11$	b. 52 is composite. $52 = 2 \times 26$	c. 41 is prime.
d. 39 is composite. $39 = 3 \times 13$	e. 43 is prime.	f. 45 is composite. $45 = 5 \times 9$

4.

number	digit sum	divisible by 3?
98	17	no
105	6	yes
567	18	yes
59	14	no

number	digit sum	divisible by 3?
888	24	yes
1,045	10	no
1,338	15	yes
612	9	yes

5.

number	divisible by 7?
99	no
74	no
56	yes

number	divisible by 7?
24	no
100	no
84	yes

number	divisible by 7?
85	no
63	yes
105	yes

6. Answers will vary, as you can write a composite number as a product in many different ways.

a. 67 is prime.	b. 57 is composite. $57 = 3 \times 19$	c. 47 is prime.
d. 53 is prime.	e. 63 is composite. $63 = 7 \times 9$	f. 61 is prime.
g. 93 is composite. $93 = 3 \times 31$	h. 85 is composite. $85 = 5 \times 17$	i. 91 is composite. $91 = 7 \times 13$
j. 87 is composite. $87 = 3 \times 29$	k. 79 is prime.	l. 97 is prime.

Finding Factors, p. 70

1.

a. factors: 1, 2, 3, 6	b. factors: 1, 2, 5, 10
c. factors: 1, 2, 3, 4, 6, 12	d. factors: 1, 3, 5, 15
e. factors: 1, 2, 4, 5, 10, 20	f. factors: 1, 2, 3, 6, 9, 18

2. Only Olivia's work was totally correct.

a. Aiden found all the factors of 34:	b. Olivia found all the factors of 28:
~~34 = 2 × 18~~ $34 = 2 \times 17$ ~~34 = 1 × 17~~ $34 = 1 \times 34$ The factors are 1, 2, 17, ~~18~~, 34	$28 = 1 \times 28$ $28 = 2 \times 14$ $28 = 4 \times 7$ The factors are 1, 2, 4, 7, 14, and 28.
c. Jayden found all the factors of 33:	d. Isabella found all the factors of 36:
$33 = 1 \times 33$ ~~33 = 3 × 13~~ $33 = 3 \times 11$ The factors are 1, 3, ~~13~~, 11, 33.	$36 = 6 \times 6$ $\underline{36 = 3 \times 12}$ $\underline{36 = 3 \times 12}$ $36 = 4 \times 9$ $\underline{36 = 1 \times 36}$ The factors are 4, 6, and 9. <u>Also 1, 2, 3, 12, 18, 36</u>

Finding Factors, continued

3.

a. factors: 1, 2, 23, 46	b. factors: 1, 2, 4, 17, 34, 68
c. factors: 1, 3, 9, 11, 33, 99	d. factors: 1, 2, 3, 4, 6, 8, 9, 12, 18, 24, 36, 72
e. factors: 1, 73	f. factors: 1, 2, 4, 5, 8, 10, 16, 20, 40, 80
g. factors: 1, 5, 19, 95	h. factors: 1, 2, 4, 8, 16, 32, 64

Mixed Review Chapter 5, p. 72

1. a. 4,284   b. 49,068

2. a. 84; 80   b. 20; 54   c. 1,090; 90

3.

Estimate:	Exact: 6,859
$1{,}568 + 4{,}839 + 452$	
$\downarrow \qquad \downarrow \qquad \downarrow$	
$\approx 1{,}600 + 4{,}800 + 500 = 6{,}900$	

4. a. 3,998; 3,960; 3,991
   b. 6,990; 9,970; 991
   c. 1,900; 6,700; 9,400

5. a. 34,268
   b. 800,046
   c. 406,780

6.

a. 3 ft = 36 in 9 ft = 108 in	b. 2 ft 5 in = 29 in 7 ft 8 in = 92 in	c. 9 ft 2 in = 110 in 10 ft 11 in = 131 in

7. a. $5{,}400 = 90 \times 60$ b. $16 \times 20 = 8 \times 40$

   c. $7 \times 49 + 49 = 8 \times 49$ d. $24{,}000 = 300 \times 80$

   e. $7 \times 13 = 5 \times 13 + 26$ f. $1{,}500 - 500 = 5 \times 200$

8. a. Estimate: 8 weeks ($8 \times \$50 = \$400$). Exact: 9 weeks, because $9 \times \$45 = \$405$. He will have $6 left over.
   b. She needs 230 cm of string, 69 sheets of paper, and 46 egg cartons.

   c. James had 25 marbles.

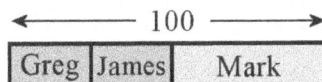

$\longleftarrow$ 100 $\longrightarrow$

Greg	James	Mark

Review Chapter 5, p. 74

1.

a.	b.	c.
$20 \div 10 + 15 = 17$ $20 \times 10 + 15 = 215$	$(200 + 100) \div 5 = 60$ $200 + 100 \div 5 = 220$	$10 \times 12 + 40 \div 10 = 124$ $10 \times (12 + 40) \div 10 = 52$

2.

a. $3{,}100 \div 100 = 31$ $450 \div 10 = 45$	b. $240 \div 20 = 12$ $800 \div 40 = 20$	c. $4{,}200 \div 600 = 7$ $3{,}200 \div 80 = 40$

3.

a.	b.	c.
45 ÷ 6 = 7 R3	12 ÷ 7 = 1 R5	31 ÷ 4 = 7 R3
46 ÷ 6 = 7 R4	27 ÷ 8 = 3 R3	56 ÷ 9 = 6 R2

4. a. 236   b. 188

5. a. 78 R2   b. 474 R1

6. 288 ÷ 4 = 72. Timmy has 72 seashells.

7. a. 70 ÷ 12 = 5 R10. Mark had five full boxes of candles.
   b. One box had ten candles.

8. $38.88 ÷ 4 = $9.72. One yard cost $9.72.

9. (92 + 85 + 89 + 75 + 89) ÷ 5 = 86. John's average score was 86.

10.

Number	13	40	57	135	354	2,380
Divisible by 3			x	x	x	
Divisible by 5		x		x		x
Divisible by 10		x				x

11.

a. Is 7 a factor of 64? No, because it does not divide evenly into 64. OR No, because 64 ÷ 7 = 9 R1; there is a remainder.	b. Is 98 a multiple of 2? Yes, because it is an even number. OR Yes, because 2 × 49 = 98.
c. Is 76 divisible by 8? No, because 76 ÷ 8 = 9 R4; the division is not even.	d. Is 30 a factor of 30? Yes, because 1 × 30 = 30.

12.

a. 87 is composite. 87 = 3 × 29	b. 89 is prime.	c. 91 is composite. 91 = 7 × 13

13.

a. factors: 1, 2, 3, 4, 6, 8, 12, 24	b. factors: 1, 3, 9, 27
c. factors: 1, 2, 3, 6, 11, 22, 33, 66	d. factors: 1, 3, 5, 15, 25, 75

Puzzle corner:
5, 11, 17, 23, 29, 35, 41, 47, 53, 59, 65, 71, 77, 83, 89, 95

Chapter 6: Geometry

Review: Area of Rectangles, p. 81

1. a. $3 \text{ cm} \times 2 \text{ cm} = 6 \text{ cm}^2$ b. $1 \text{ ft} \times 1 \text{ ft} = 1 \text{ ft}^2$ c. $4 \text{ km} \times 8 \text{ km} = 32 \text{ km}^2$

2. a. 5 in   b. 10 cm   c. 25 ft

3.

a. $12 \text{ cm} \times s = 96 \text{ cm}^2$ $s = 8 \text{ cm}$	b. $20 \text{ cm} \times s = 1{,}000 \text{ cm}^2$ $s = 50 \text{ cm}$

4. a. $9 \text{ m} \times s = 45 \text{ m}^2$ $s = 5 \text{ m}$
   b. $60 \text{ ft} \times s = 1{,}800 \text{ ft}^2$ $s = 30 \text{ ft}$

5. a. $36 \text{ in} \times 24 \text{ in} = 864 \text{ in}^2$ (the student needs to use the multiplication algorithm with pencil and paper)
   b. $3 \text{ ft} \times 2 \text{ ft} = 6 \text{ ft}^2$

6. a. $4 \times (2 + 5) = 4 \times 2 + 4 \times 5 = 8 + 20 = 28$
   b. $3 \times (5 + 2) = 3 \times 5 + 3 \times 2 = 15 + 6 = 21$

7. a. $A = 30 \text{ ft} \times (30 \text{ ft} + 75 \text{ ft})$
       $= 30 \text{ ft} \times 30 \text{ ft} \ + \ 30 \text{ ft} \times 75 \text{ ft}$
       $= 900 \text{ ft}^2 + 2{,}250 \text{ ft}^2 \ = \ 3{,}150 \text{ ft}^2$

   b. $2{,}250 \text{ ft}^2 - 900 \text{ ft}^2 = 1350 \text{ ft}^2$

8. Check the student's drawings.   a. 2 in.   b. 3 cm   c. 1 ft

Problem Solving: Area of Rectangles, p. 84

1. a. There are several possible number sentences, depending on how you divide the shape into rectangles.
      $4 \times 5 + 3 \times 2 + 2 \times 6 = 20 + 6 + 12 = 38$ square units
      or $4 \times 2 + 8 \times 3 + 2 \times 3 = 8 + 24 + 6 = 38$ square units

   b. Here it is handy to use subtraction. $7 \times 6 - 3 \times 2 = 36$ square units

2. a.

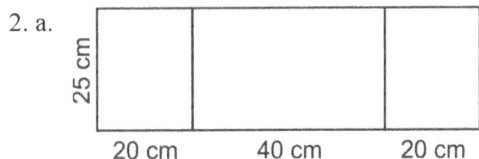

   b. $25 \text{ cm} \times (20 \text{ cm} + 40 \text{ cm} + 20 \text{ cm}) = 25 \text{ cm} \times 80 \text{ cm} = 2{,}000 \text{ cm}^2$.

3. Answers will vary. Check the student's answers. For example:

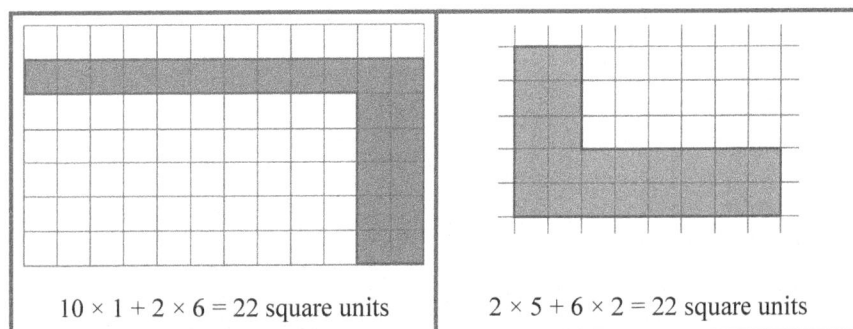

$10 \times 1 + 2 \times 6 = 22$ square units $2 \times 5 + 6 \times 2 = 22$ square units

4. The area is $48 \text{ ft} \times 48 \text{ ft} \ - \ 30 \text{ ft} \times 20 \text{ ft} = 2304 \text{ ft}^2 - 600 \text{ ft}^2 = 1704 \text{ ft}^2$.

5. a. $3 \text{ m} \times 3 \text{m} \ + \ 7 \text{ m} \times 5 \text{ m} \ + \ 3 \text{ m} \times 3 \text{ m} = 9 \text{ m}^2 + 35 \text{ m}^2 + 9 \text{ m}^2 = 53 \text{ m}^2$.
   b. Divide the shape into two rectangles. That can be done in two ways. One way results in the calculation
      $16 \text{ ft} \times 32 \text{ ft} \ + \ 24 \text{ ft} \times 12 \text{ ft} = 512 \text{ ft}^2 + 288 \text{ ft}^2 = 800 \text{ ft}^2$.

1. a. perimeter   b. volume   c. area

2. a. perimeter = 32 in   area = 48 in^2
   b. perimeter = 200 cm   area = 2,500 cm^2

3. a. perimeter = 16 cm   area = 7 cm^2
   b. perimeter = 32 cm   area = 28 cm^2. Notice that when the perimeter gets doubled, the area gets *quadrupled*.

4. You can also use a letter or other symbol for the unknown, instead of $\underline{?}$.

   a. 9 ft + $\underline{?}$ + 9 ft + $\underline{?}$ = 32 ft. Solution: $\underline{?}$ = 7 ft.

   b. 15 cm + $\underline{?}$ + 15 cm + $\underline{?}$ = 44 cm or 15 cm + $\underline{?}$ = 22 cm. Solution: $\underline{?}$ = 7 cm.

   c. $\underline{?}$ + $\underline{?}$ + $\underline{?}$ + $\underline{?}$ = 24 m or $\underline{?}$ + $\underline{?}$ = 12 m or 4 × $\underline{?}$ = 24 m. Solution: $\underline{?}$ = 6 m.

   d. 5 cm. The number sentence would be $s \times s = 25$ cm^2. Its perimeter is 20 cm.

5.

a. Sides 4 and <u>5</u> units; area 20 square units, perimeter <u>18</u> units.	b. Sides <u>3</u> and <u>5</u> units; area 15 square units, perimeter 16 units.	c. Sides 3 and <u>7</u> units; area <u>21</u> square units, perimeter 20 units.

6. 516 mm ÷ 6 = 86 mm

7. First divide the building into two rectangles. That can be done in two different ways. One way is shown on the right.
   a. The number sentence is then: 18 m × 33 m + 42 m × 18 m.

   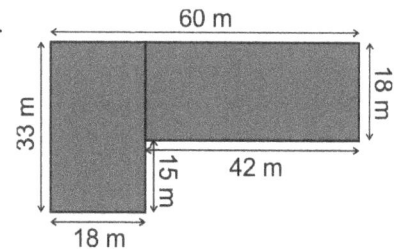

   b. The area is 18 m × 33 m + 42 m × 18 m
      = 594 m^2 + 756 m^2 = 1,350 m^2.

   c. The area of the whole plot of land is 80 m × 35 m = 2,800 m^2.
      We subtract the area of the school to get the area of the actual yard:
      2,800 m^2 − 1,350 m^2 = 1,450 m^2.

8. a. & b.

	Length	Width	Area	Fencing needed
Sheep yard 1	20 m	20 m	400 m^2	80 m
Sheep yard 2	10 m	40 m	400 m^2	100 m
Sheep yard 3	5 m	80 m	400 m^2	170 m

9. For example, 5 m by 10 m pen will work. Or, 6 m by 9 m. Or, 7 m by 8 m.

Puzzle corner. The width of the inside rectangle is 19 cm − 2 cm − 2 cm = 15 cm. The height of the inner rectangle is 14 cm − 2 cm − 2 cm = 10 cm. The area is 15 cm × 10 cm = 150 cm^2.

Lines, Rays, and Angles, p. 90

1. a. Ray $\overrightarrow{AB}$ b. line $\overleftrightarrow{VW}$ c. line segment $\overline{MN}$

2.

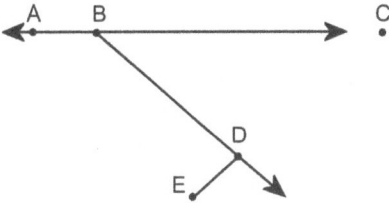

3. angle MAS or angle SAM

4. a. angle EDF or angle FDE
   b. angle ACE or angle ECA

5. Answers will vary. Check students' answers. For example:

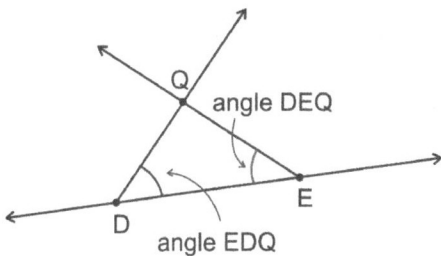

6. a. The second angle is bigger.   b. the second angle   c. the first angle
   d. the second angle   e. the first angle   f. the second angle

Measuring Angles, p. 93

1. a. 35°   b. 72°   c. 18°   d. 50°

2. a. 75°   b. 100°
   c. 144°   d. 135°
   e. 173°   f. 93°

3. It is 180° −146° = 34°.

4. a. 70°   b. 45°   c. 148°   d. 125°   e. 76°   f. 107°   g. 14°

5. a. acute   b. right   c. straight   d. right   e. acute   f. obtuse   g. acute   h. acute   i. obtuse

6. Check students' work. For example:

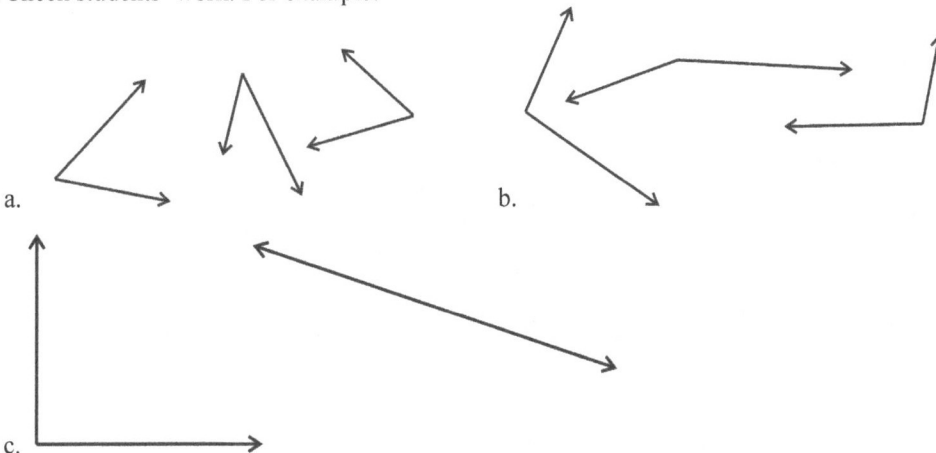

a.

b.

c.

7. The three angles measure 45° (acute), 102° (obtuse), 33° (acute).

Drawing Angles, p. 100

1. Check the student's answers.

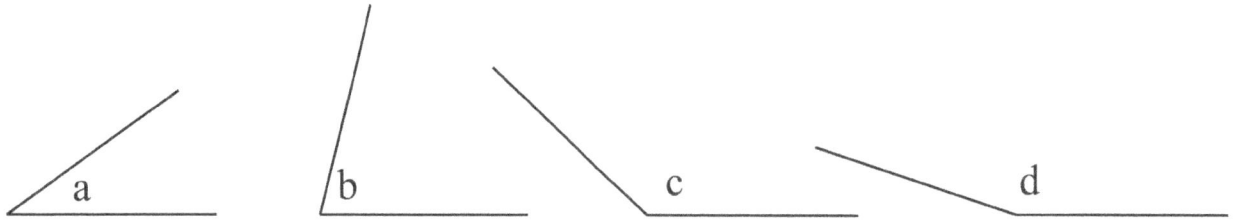

2. This is the general shape of the triangle
   with 85 and 25-degree angles:

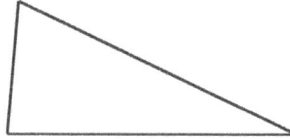

3. This is the general shape of the triangle with 30 and 60-degree
   angles. The student's triangle may also be a mirror
   image of this. Note: the third angle will be a right angle.

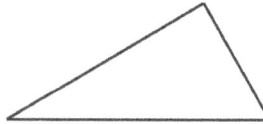

Estimating Angles, p. 102

1. a. 60°   b. 150°

2. Answers will vary. The following are the exact measurements.
   a. 30°   b. 45°   c. 100°
   d. 60°   e. 160°   f. 80°

3. Answers will vary. The following are the exact measurements.
   a. 40°   b. 10°   c. 15°
   d. 120°   e. 270°   f. 210°

4. Answers will vary. The following are the exact measurements: 60°, 30°, 90°.

5. Answers will vary. Check the student's answers.

6.

a. The drum

b. The caterpillar

c. The train

d. The top

7.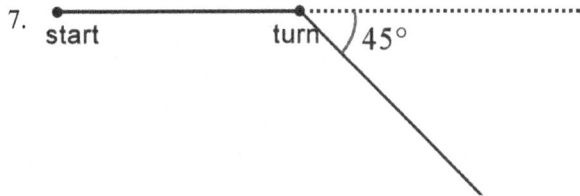

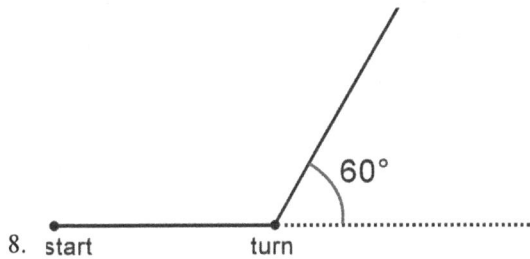

8.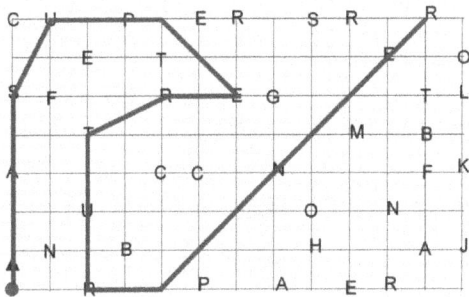

Puzzle corner: A SUPER TURNER

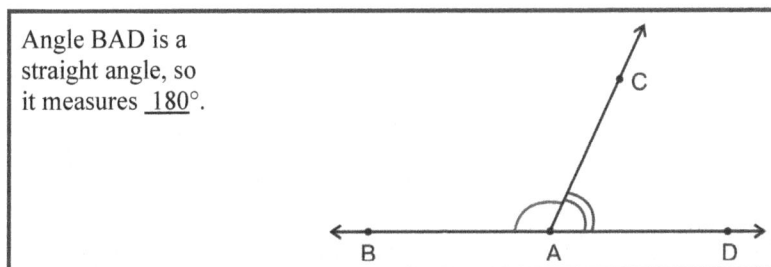

Angle Problems, p. 107

1. Yes. You can subtract the angle measure of ∠DBC from 90° to get the measure of ∠ABD.

2. a. 74°   b. 58°   c. 65°   d. 63°

3. a. 35°   b. 45°

4. a. 38° + x = 72°; x = 34°
   b. 37° + x = 90°; x = 53°
   c. 47° + 23° + x = 122°; x = 52°
   d. 34° + x = 105°; x = 71°

Angle BAD is a
straight angle, so
it measures _180°_.

```
              C
             ↗
            /
  ⟵———————————————⟶
    B       A       D
```

---

5. a. 68°   b. 88°

6. a. 29° + x = 180°; x = 151°
   b. x + 54° = 180°; x = 126°

Angle Problems, continued

There are THREE angles in this picture that share the same vertex. Notice that they three make a complete circle!

Measure them. Find the sum of the angle measures.

Angle A: _155°_

Angle B: _80°_

Angle C: _125°_

Sum: _360°_

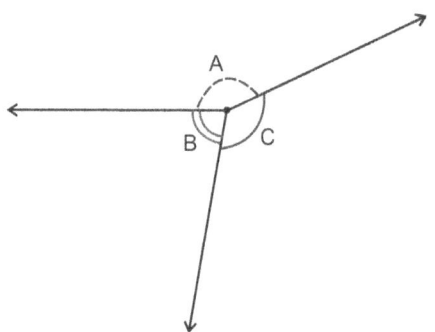

7. a. $28° + 116° + 54° + x = 360°$; $x = 162°$
   b. $57° + 113° + x = 360°$; $x = 190°$.

8. The sum of all three angle measures is 300°. (The first angle measures 180°, the second 90°, and the last one 30°.)

9. Measure first the smaller angle that you can measure with the protractor, marked with a dashed line. It is 102°. Then subtract that from 360°: $360° − 102° = 258°$. That is the measurement of the angle in question.

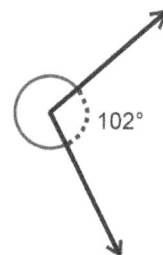

Puzzle corner. a. It is $360° ÷ 8 = 45°$.   b. 120°. (First divide 360° by six, then double that.)

Parallel and Perpendicular Lines, p. 112

1. They are not parallel. They intersect.

2. a. Line segments AB and BC are _perpendicular_.
      Line segments AD and BC are _parallel_.

   b. Line segments EF and GH are _parallel_.
      Line segments EH and FG are _parallel_.

3.

4.
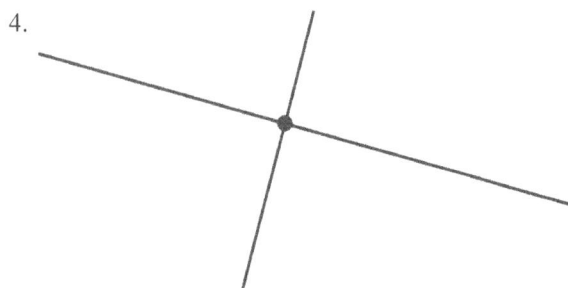

5.
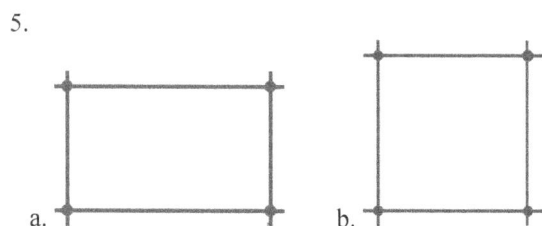

6. Answers will vary. Check students' answers.

7. Answers will vary. For example:

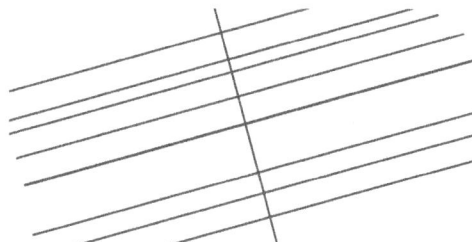

8. Answers will vary. Check the student's answers. For example:

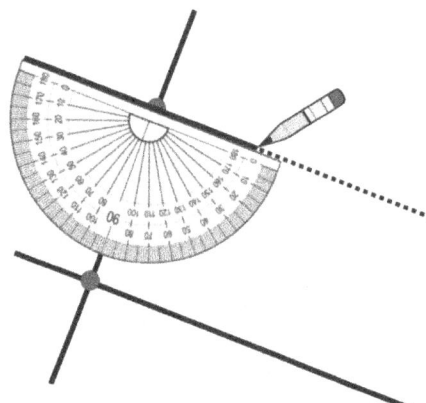

9. $\overline{AF} \perp \overline{FB}$ $\overline{AF} \parallel \overline{BE}$

   $\overline{FB} \perp \overline{BE}$ $\overline{AB} \parallel \overline{FE}$

Parallel and Perpendicular Lines, cont.

10. a. $\overline{AB} \perp \overline{AD}$, $\overline{AD} \perp \overline{CD}$, $\overline{AB} \parallel \overline{CD}$

b. $\overline{AB} \perp s$, $r \perp t$, $t \perp u$, $r \parallel u$

c. $s \perp t$, $s \perp \overline{BC}$, $s \perp \overline{FE}$,
$\overline{AF} \parallel \overline{CD}$, $\overline{AB} \parallel \overline{DE}$, $\overline{BC} \parallel \overline{EF}$, $\overline{EF} \parallel t$, $\overline{BC} \parallel t$.

Parallelograms, p. 117

1. a. Answers will vary. For example:

b. Answers will vary, but the opposite sides should measure the same.

2. Answers will vary. Check the student's answers. The opposite sides should measure the same (be congruent), as also should the opposite corners.

3. Answers will vary. Check the student's answers. For example:

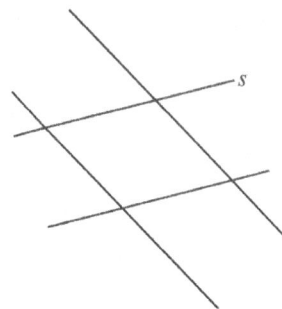

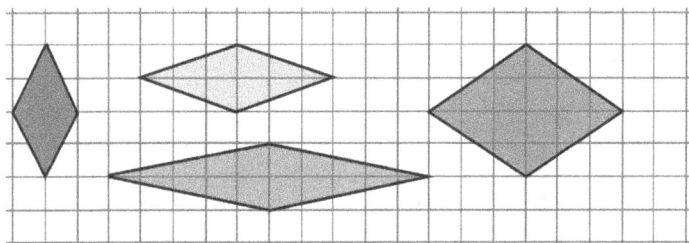

4. Answers will vary. Check the student's answers. For example:

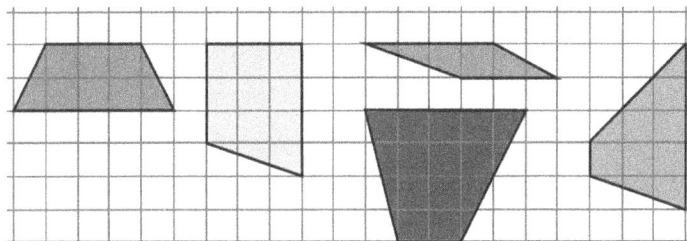

5. a. trapezoid   b. parallelogram   c. trapezoid   d. rhombus

6. a. The image is not to scale:

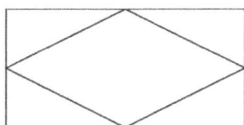

b. a rhombus. (It is also a parallelogram.)

7. A: trapezoid   B: parallelogram   C: trapezoid
D: rectangle   E: rhombus   F: (no special name or a *scalene quadrilateral*)

Puzzle corner:
Yes. A rectangle is a parallelogram, because its opposite sides are parallel.
A square is a parallelogram, too.
A square is a rhombus, yes.
The rectangle in the question is not a rhombus.

Triangles, p. 120

1. a. and b. Use a protractor or a triangular ruler to draw the right angle.
   c. The other two angles in a right triangle are acute.

> **A right triangle has one right angle.**
> **The other two angles are <u>acute</u>.**

2. c. In an obtuse triangle the other two angles are acute.

> **An obtuse triangle has one obtuse angle.**
> **The other two angles are <u>acute</u>.**

3. a. b. Answers will vary. Check the student's answers. All the angles should be acute (less than 90°).

4. **Right triangles** have exactly one <u>right angle</u>, and the other two angles are <u>acute</u>.

   Obtuse triangles have exactly one <u>obtuse angle</u>, and the other two angles are <u>acute</u>.

   Acute triangles have <u>three</u> <u>acute</u> angles.

5. a. obtuse
   b. acute
   c. acute
   d. obtuse
   e. acute
   f. right
   g. Triangle ABD: right.
      Triangle ACD: obtuse.
      Triangle BCD: right.

6. a. acute
   b. right
   c. obtuse
   d. acute
   e. The black triangle is obtuse. The red triangle is acute.

7. a.
   b. The measure of the third angle is 105°.
   c. It is an obtuse triangle.

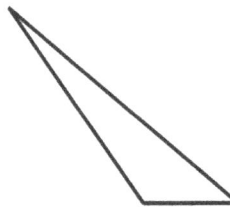

8. a.
   b. The measure of the third angle is 15°.
   c. It is an obtuse triangle.

9. a.
   b. The measure of the third angle is 90°.
   c. It is a right triangle.

Line Symmetry, p. 124

1. a. no   b. yes   c. yes   d. no   e. no   f. no   g. no   h. yes   i. yes

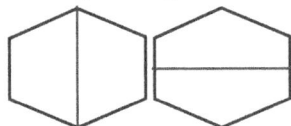

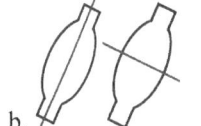

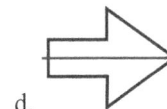

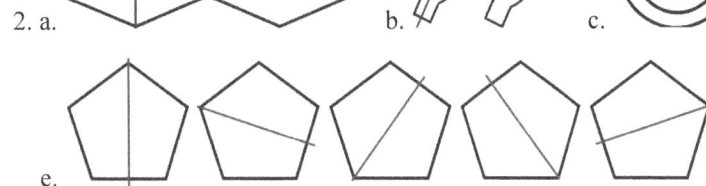

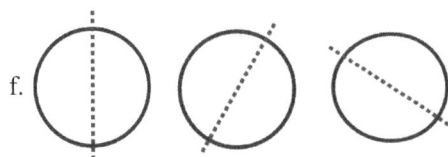

2. a.   b.   c.   d.

   e.

   f.

   ...and many more. Any diameter of a circle (a line through its center point) is its symmetry line.

3. You can draw a vertical symmetry line in the letters A, H, I, M, O, T, U, V, W, X, and Y.
   You can draw a horizontal symmetry line in the letters B, C, D, E, H, I, K, O, and X.

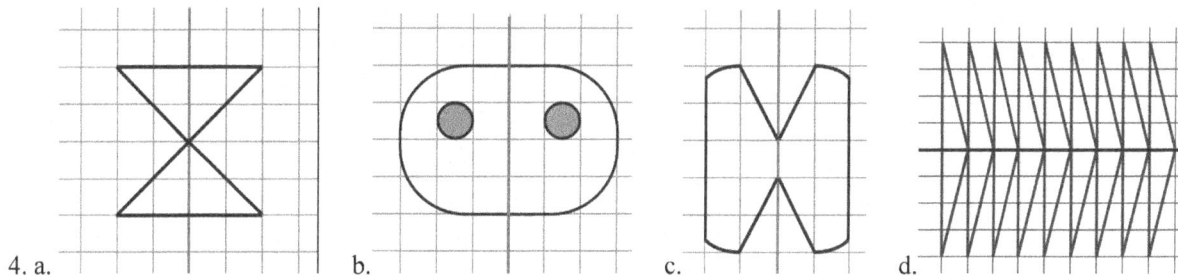

4. a.   b.   c.   d.

Mixed Review Chapter 6, p. 127

1. a. 3,000 g;   7,400 g
   b. 5,000 ml;   2,060 ml
   c. 9,000 m;   4,250 m

2. There are 1 1/2 liters left now. Six glasses got filled.

3. a. 48 in;   74 in
   b. 24 fl. oz.;   8 qt
   c. 64 oz;   121 oz

4. The 1-pint bottle is more. It is 4 ounces more.

5. a. 8 in
   b. 24 in or 2 ft
   c. 8 ft

6. a. 960   b. 508   c. 1,670   d. 1,099

7.

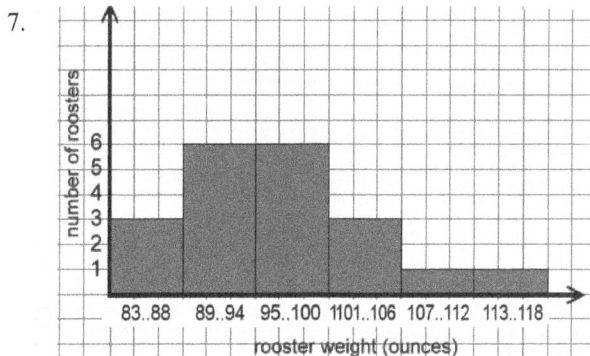

Weight (ounces)	Frequency
83..88	3
89..94	6
95..100	6
101..106	3
107..112	1
113..118	1

8. a. The silverware set costs $4 × $13 = $52. The two items together cost $13 + $52 = $65.
   b. He still has $200 − 8 × $18 = $56.
   c. From 22:15 p.m. till 7:00 a.m. is 8 hours 45 minutes. But he did not sleep from 3:30 till 5:10, which is 1 hour 40 minutes. So, we subtract those two mounts and get that he slept 8 h 45 min − 1 h 40 min = 7 h 5 min.

1. Side 1: 17 1/2 ft; Side 2: 10 ft; Side 3: 17 1/2 ft

2. The area is 320 m².
   First, divide the shape into two rectangles. You need to use subtraction to find some missing lengths of sides.

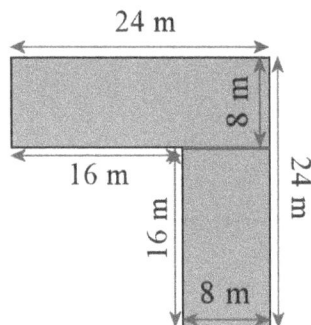

The upper rectangle is 24 m × 8 m so its area is 192 m².
The lower rectangle is 8 m × 16 m so its area is 128 m².
In total, the area is 320 m².

3. a. 100°   b. 33°

4. a. right   b. acute   c. obtuse   d. acute   e. acute
   f. right   g. right   h. obtuse   i. right   j. obtuse
   k. obtuse

5. Check the student's answer. Here is one such angle:

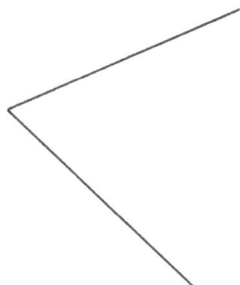

6. It is 84°. The student can figure it out in any manner.
   If we use an equation, we write:
   64° + x + 32° = 180°;   x = 84°.

7.

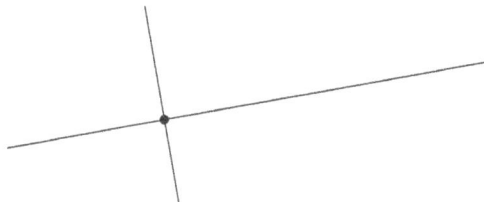

8. Task b. is possible to do. Answers will vary as the other two angles can vary. For example:

9. Check the student's drawing.

10. They are right triangles.

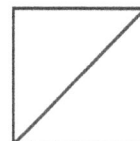

11. $\overline{AB} \parallel m$, $\overline{BC} \parallel n$,
    $\overline{AC} \perp m$, $\overline{AB} \perp \overline{AC}$

12. Answers will vary. Check the student's drawing.

13. a.

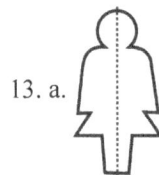

    b. not symmetrical

    c.

    d. not symmetrical

    e.

84

Chapter 7: Fractions

One Whole and Its Fractional Parts, p. 137

1.

a. Color 1 part.	b. Color 5 parts.	c. Color 8 parts.	d. Color 3 parts.
$\frac{1}{12}$ and $\frac{11}{12}$	$\frac{5}{10}$ and $\frac{5}{10}$	$\frac{8}{9}$ and $\frac{1}{9}$	$\frac{3}{7}$ and $\frac{4}{7}$

2.

a. $1 = \frac{9}{9}$	b. $1 = \frac{3}{3}$	c. $1 = \frac{12}{12}$	d. $1 = \frac{4}{4}$	e. $1 = \frac{5}{5}$

3. a. 1/4 of the pie is left.   b. 5/6 of the pizza is left.   c. 1/5 of the bar.

4. a. 2/5   b. 4/5   c. 1 4/5   d. 2 3/5   e. 3 1/5   f. 4 2/5   g. 5 3/5

5. a. 2/6   b. 5/6   c. 1 1/6   d. 1 4/6   e. 2 3/6   f. 2 5/6

6. a. 1/3   b. 1 1/3   c. 1 2/3   d. 2 2/3

7.

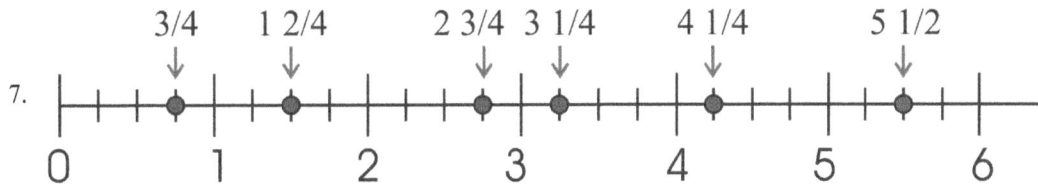

8.

a. Color 1 part.	b. Color 10 parts.	c. Color 3 parts.	d. Color 15 parts.
$\frac{1}{6} + \frac{5}{6} = 1$	$\frac{10}{12} + \frac{2}{12} = 1$	$\frac{3}{8} + \frac{5}{8} = 1$	$\frac{15}{100} + \frac{85}{100} = 1$

9. a. 1/4   b. 1/7   c. 7/8   d. 1/12

10. a. 2/4 or 1/2 liter
    b. 14/20 of the bread is left.

11. a. 1/10 of 90 km = 9 km. Then, 4/10 of 90 km = 36 km.
    b. First, divide $45.50 into five parts: $45.50 ÷ 5 = $9.10. Cindy pays 2/5 of the bill, or double that, which is $18.20.
       Sandy pays the rest, or $27.30.
    c. 7/9 is left. $2,100 is left. One-ninth of his paycheck is $300, so seven-ninths of it is 7 × $300 = $2,100.

Mixed Numbers, p. 140

1. a. 2 3/4   b. 1 1/2   c. 4 2/10
   d. 8 1/3   e. 2 4/9   f. 3 5/6

2.

a. $2\frac{2}{4} + \frac{2}{4} = 3$	b. $2\frac{2}{6} + \frac{4}{6} = 3$	c. $1\frac{1}{5} + \frac{4}{5} = 2$
d. $5\frac{1}{3} + \frac{2}{3} = 6$	e. $3\frac{3}{4} + \frac{1}{4} = 4$	f. $1\frac{3}{6} + \frac{3}{6} = 2$

3. a. 3/4   b. 8/10   c. 5/9   d. 7/8

4.

a. 2 = 6/3	b. 3 = 6/2	c. 3 = 12/4

5.

a. $3 = \frac{12}{4}$	b. $1 = \frac{9}{9}$	c. $4 = \frac{4}{1}$	d. $7 = \frac{35}{5}$	e. $6 = \frac{60}{10}$
f. $7 = \frac{7}{1}$	g. $10 = \frac{60}{6}$	h. $20 = \frac{60}{3}$	i. $24 = \frac{48}{2}$	j. $50 = \frac{250}{5}$

6.

a. $1 = \frac{2}{2} = \frac{4}{4} = \frac{7}{7} = \frac{9}{9} = \frac{20}{20}$	b. $4 = \frac{4}{1} = \frac{20}{5} = \frac{40}{10} = \frac{44}{11} = \frac{120}{30}$

7. a. 3   b. 9   c. 30   d. 9   e. 30

8. Answers will vary. For example:
   a. 4/5 = 1/5 + 1/5 + 2/5 or 1/5 + 3/5 or 2/5 + 2/5
   b. 5/8 = 1/8 + 1/8 + 3/8 or 2/8 + 3/8 or 2/8 + 2/8 + 1/8
   c. 9/12 = 3/12 + 3/12 + 3/12 or 2/12 + 5/12 + 2/12 or 1/12+ 4/12 + 4/12
   d. 4/3 = 2/3 + 2/3 or 1/3 + 3/3
   e. 9/6 = 1/6 + 8/6 or 2/6 + 7/6 or 3/6 + 6/6

9. She can still pour 3/4 cup of water into the pitcher.

10. There is two-thirds of a pound of extra beef.

11. He had 1 7/12 of the bread left.

12. Your train of cars would be 4 1/2 inches long.

13. She needs five scoops of flour.

Puzzle corner.
a. It is not correct. You could change the total to 3 2/4 to make it correct, or change one of the addends
   to be 1/4 less than in the problem.
b. It is correct.

86

Mixed Numbers and Fractions, p. 144

1. a. 1 7/8 b. 1 2/3 c. 2 1/5
   d. 2 3/4 e. 3 4/6 f. 3 1/2

2. a. 2 3/5 b. 3 2/3 c. 5 3/4 d. 8 1/2
   e. 3 5/7 f. 6 1/9 g. 2 2/10 h. 7 2/8

3. a. 13/9 b. 8/5 c. 21/8

4. a. 12/5 b. 4/3 c. 13/4 d. 9/2
   e. 21/4 f. 19/3 g. 26/3 h. 81/10

5.

$2\frac{2}{5}$	$1\frac{4}{5}$	$4\frac{7}{8}$	$3\frac{1}{5}$	$3\frac{3}{5}$	$5\frac{5}{8}$	$4\frac{1}{8}$
$\frac{45}{8}$	$\frac{12}{5}$	$\frac{16}{5}$	$\frac{9}{5}$	$\frac{33}{8}$	$\frac{18}{5}$	$\frac{39}{8}$

6. a. 7 3/6 b. 22/3 c. 2 7/20 d. 23/6
   e. 5 3/4 f. 22/5 g. 4 6/7 h. 43/4

7. Answers will vary. For example:

   a. 2 5/7 = 1 + 6/7 + 6/7
      2 5/7 = 1 2/7 + 5/7 + 5/7
      2 5/7 = 4/7 + 10/7 + 5/7

   b. 2 = 5/6 + 3/6 + 4/6
      2 = 1 2/6 + 1/6 + 3/6
      2 = 4/6 + 1 1/6 + 1/6

8. The library has 2,610 ÷ 9 × 7 = 2,030 children's fiction books.
   And it has 2,610 − 2,030 = 580 children's nonfiction books.

9. a. 11/12 b. 7/9 c. 11/15 d. 13/20

10. a. 16/2 b. 70/7 c. 66/11 d. 80/4 e. 96/8

Adding Fractions, p. 147

1.

a. $\frac{1}{6} + \frac{3}{6} = \frac{4}{6}$	b. $\frac{2}{8} + \frac{5}{8} = \frac{7}{8}$
c. $\frac{7}{8} + \frac{7}{8} = \frac{14}{8} = 1\frac{6}{8}$	d. $\frac{7}{10} + \frac{5}{10} = \frac{12}{10} = 1\frac{2}{10}$

Adding Fractions, cont.

2.

a. $\dfrac{3}{5} + \dfrac{4}{5} = 1\dfrac{2}{5}$

b. $1\dfrac{2}{5} + \dfrac{4}{5} = 2\dfrac{1}{5}$

c. $\dfrac{13}{10} + \dfrac{6}{10} = 1\dfrac{9}{10}$

d. $1\dfrac{3}{8} + \dfrac{6}{8} = 2\dfrac{1}{8}$

3. a. 2/6   b. 1   c. 7/8
   d. 1 2/5   e. 2
   f. 1 8/10   g. 1 2/4

4. 3/12 + 2/12 + 4/12 = 9/12. The children ate 9/12 of the chocolate bar. There is 3/12 left.

5. a. 2 2/5   b. 2 6/8   c. 2

6. a. 9/12   b. 3/6   c. 4/8

Adding Mixed Numbers, p. 149

1.

a. $1\dfrac{3}{5} + 2\dfrac{2}{5} = 4$

b. $1\dfrac{3}{7} + 2\dfrac{6}{7} = 4\dfrac{2}{7}$

c. $1\dfrac{3}{8} + 1\dfrac{6}{8} = 3\dfrac{1}{8}$

d. $\dfrac{8}{9} + 1\dfrac{5}{9} = 2\dfrac{4}{9}$

2. a. 5   b. 7 1/6
   c. 13 1/4   d. 10 2/8
   e. 27 2/6   f. 12 2/10

3. These answers can be fixed in different ways. For example:

a. In the first one, Emma has one seventh too much. The second one is correct.	b. In the first one, Peter is lacking one third from the addition. In the second one, he has one third too much.
$1\dfrac{5}{7} = \dfrac{2}{7} + 1\dfrac{1}{7} + \dfrac{2}{7}$	$2\dfrac{1}{3} = \dfrac{2}{3} + \dfrac{2}{3} + \dfrac{2}{3} + \dfrac{1}{3}$
$1\dfrac{5}{7} = \dfrac{10}{7} + \dfrac{2}{7}$	$2\dfrac{1}{3} = \dfrac{5}{3} + \dfrac{2}{3}$

4. Answers will vary. For example:
   a. 1 3/10 = 1 + 3/10 or 5/10 + 8/10 or 4/10 + 9/10 or 2/10 + 11/10

   b. 3 1/5 = 4/5 + 12/5 or 10/5 + 6/5   or 8/5 + 8/5 or 1/5 + 15/5

Adding Mixed Numbers, cont.

5. a. 1 1/2 + 1/2 + 1/2 = 2 1/2. The recipe calls for 2 1/2 cups of flour.
   b. 1 3/4 + 1 1/4 = 3. They took three hours.
   c. 1 1/2 + 3/4 = 2. He drank 2 cups of liquid.

6. 2 1/4 + 3 1/4 + 2 1/4 + 3 1/4 = 11. Its perimeter is 11 inches.

7. 2 3/8 + 2 3/8 + 2 3/8 = 6 9/8 = 7 1/8. Its perimeter is 7 1/8 inches.

8. For the amount of sugar, 1 2/4 is also acceptable, and for the amount of flour, 2 2/4 cups is also acceptable, as students have not yet been taught how to simplify fractions.

> <u>A birthday cake</u>
> 8 eggs
> 1 1/2 (or 1 2/4) cups sugar
> 2 1/2 (or 2 2/4) cups flour
> 3 tsp baking powder
> 2 cups whipped cream
> sliced fruit

Puzzle corner:
a. 1 1/5   b. 1 3/4   c. 1 3/6

Equivalent Fractions, p. 152

1.

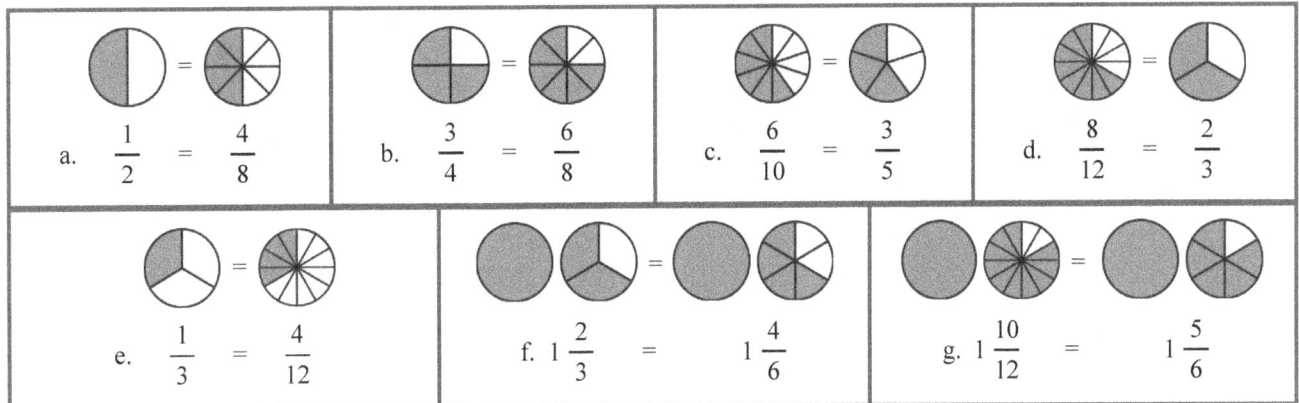

a. $\frac{1}{2} = \frac{4}{8}$ b. $\frac{3}{4} = \frac{6}{8}$ c. $\frac{6}{10} = \frac{3}{5}$ d. $\frac{8}{12} = \frac{2}{3}$

e. $\frac{1}{3} = \frac{4}{12}$ f. $1\frac{2}{3} = 1\frac{4}{6}$ g. $1\frac{10}{12} = 1\frac{5}{6}$

2.

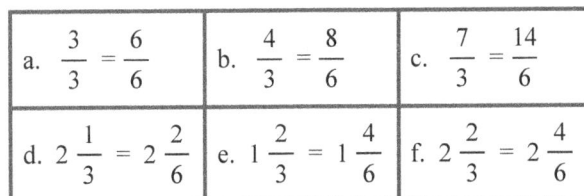

a. $\frac{3}{3} = \frac{6}{6}$ b. $\frac{4}{3} = \frac{8}{6}$ c. $\frac{7}{3} = \frac{14}{6}$

d. $2\frac{1}{3} = 2\frac{2}{6}$ e. $1\frac{2}{3} = 1\frac{4}{6}$ f. $2\frac{2}{3} = 2\frac{4}{6}$

3.

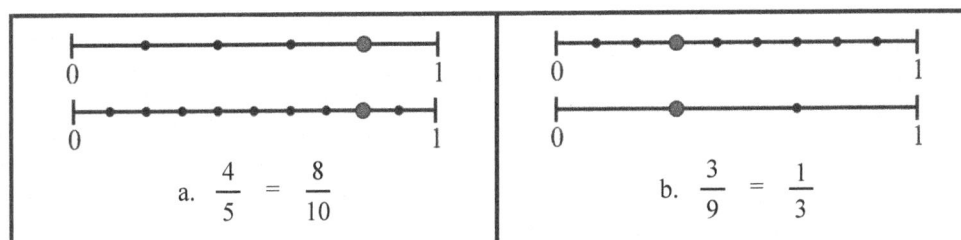

a. $\frac{4}{5} = \frac{8}{10}$ b. $\frac{3}{9} = \frac{1}{3}$

4. a.

$$\frac{1}{2} = \frac{2}{4}$$

b.

$$\frac{1}{2} = \frac{4}{8}$$

c.

$$\frac{1}{4} = \frac{3}{12}$$

d.

$$\frac{1}{3} = \frac{3}{9}$$

e.

$$\frac{5}{6} = \frac{10}{12}$$

f.

$$\frac{2}{5} = \frac{6}{15}$$

Can you notice a *shortcut* for finding the second fraction without using a picture?

g.

$$\frac{2}{3} = \frac{8}{12}$$

h.

$$\frac{4}{5} = \frac{8}{10}$$

i.

$$\frac{3}{4} = \frac{9}{12}$$

_**If**_ **you found the shortcut, explain how it works in these problems:**

$$\frac{1}{3} = \frac{3}{9}$$

Multiply the top and bottom numbers by 3.

$$\frac{3}{5} = \frac{6}{10}$$

Multiply the top and bottom numbers by 2.

5.

a. Each piece is split into __3__ new ones.

$$\frac{3}{4} \overset{\times\ \underline{3}}{\underset{\times\ \underline{3}}{=}} \frac{9}{12}$$

b. Each piece is split into __4__ new ones.

$$\frac{1}{3} \overset{\times\ \underline{4}}{\underset{\times\ \underline{4}}{=}} \frac{4}{12}$$

c. Each piece is split into __5__ new ones.

$$\frac{1}{2} \overset{\times\ \underline{5}}{\underset{\times\ \underline{5}}{=}} \frac{5}{10}$$

d.

$$\frac{1}{4} \overset{\times\ \underline{4}}{\underset{\times\ \underline{4}}{=}} \frac{4}{16}$$

e.

$$\frac{2}{3} \overset{\times\ \underline{3}}{\underset{\times\ \underline{3}}{=}} \frac{6}{9}$$

f.

$$\frac{2}{3} \overset{\times\ \underline{4}}{\underset{\times\ \underline{4}}{=}} \frac{8}{12}$$

g.

$$\frac{4}{5} \overset{\times\ \underline{2}}{\underset{\times\ \underline{2}}{=}} \frac{8}{10}$$

h.

$$\frac{2}{3} \overset{\times\ \underline{5}}{\underset{\times\ \underline{5}}{=}} \frac{10}{15}$$

i.

$$\frac{2}{5} \overset{\times\ \underline{3}}{\underset{\times\ \underline{3}}{=}} \frac{6}{15}$$

6. a. 15/18   b. 15/20   c. 8/20   d. 90/100

7.

a. Pieces were split into 3 new ones.	b. Pieces were split into 10 new ones.	c. Pieces were split into 6 new ones.	d. Pieces were split into 5 new ones.
$\dfrac{1}{2} = \dfrac{3}{6}$	$\dfrac{3}{10} = \dfrac{30}{100}$	$\dfrac{2}{5} = \dfrac{12}{30}$	$\dfrac{7}{8} = \dfrac{35}{40}$
e. $\dfrac{2}{3} = \dfrac{4}{6}$	f. $\dfrac{3}{5} = \dfrac{9}{15}$	g. $\dfrac{5}{6} = \dfrac{10}{12}$	h. $\dfrac{1}{3} = \dfrac{3}{9}$

8.

a. $\dfrac{1}{10} = \dfrac{10}{100}$	b. $\dfrac{3}{10} = \dfrac{30}{100}$	c. $\dfrac{6}{10} = \dfrac{60}{100}$	d. $\dfrac{4}{10} = \dfrac{40}{100}$	e. $\dfrac{13}{10} = \dfrac{130}{100}$

9.

10.

a. $\dfrac{1}{2} = \dfrac{2}{4} = \dfrac{3}{6} = \dfrac{4}{8} = \dfrac{5}{10} = \dfrac{6}{12} = \dfrac{7}{14}$	b. $\dfrac{1}{3} = \dfrac{2}{6} = \dfrac{3}{9} = \dfrac{4}{12} = \dfrac{5}{15}$

11. a. 18/100   b. 73/100   c. 75/100
    d. 99/100   e. 93/100   f. 114/100 or 1 14/100
    g. 147/100 or 1 47/100   h. 3 78/100   i. 102/100 or 1 2/100

12. Answers will vary. For example:

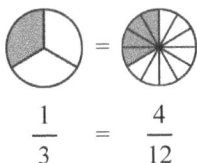

$$\frac{1}{3} = \frac{4}{12}$$

Puzzle corner:

a. $\dfrac{3}{4} + \dfrac{1}{2}$	b. $\dfrac{1}{5} + \dfrac{3}{10}$	c. $\dfrac{2}{3} + \dfrac{2}{9}$
↓ ↓	↓ ↓	↓ ↓
$\dfrac{3}{4} + \dfrac{2}{4} = \dfrac{5}{4} = 1\dfrac{1}{4}$	$\dfrac{2}{10} + \dfrac{3}{10} = \dfrac{5}{10}$	$\dfrac{6}{9} + \dfrac{2}{9} = \dfrac{8}{9}$

Subtracting Fractions and Mixed Numbers, p. 157

1. a. 8/10   b. 4/12   c. 2 2/6
   d. 1 2/9   e. 6/4 = 1 2/4   f. 2 4/8
   g. 3 4/12   h. 2/10   i. 3 4/12   j 1/8

2. a. 3/6   b. 6/10   c. 3/8   d. 3/5

3.

a. $3\frac{2}{10} - \frac{6}{10}$ $\downarrow \quad \downarrow$ $2\frac{12}{10} - \frac{6}{10} = 2\frac{6}{10}$	b. $2\frac{1}{7} - \frac{5}{7}$ $\downarrow \quad \downarrow$ $1\frac{8}{7} - \frac{5}{7} = 1\frac{3}{7}$	c. $5\frac{3}{9} - 2\frac{7}{9}$ $\downarrow \quad \downarrow$ $4\frac{12}{9} - 2\frac{7}{9} = 2\frac{5}{9}$	d. $7\frac{2}{5} - 4\frac{4}{5}$ $\downarrow \quad \downarrow$ $6\frac{7}{5} - 4\frac{4}{5} = 2\frac{3}{5}$

4. 6 1/4   b. 5 1/9   c. 3 1/2   d. 7 2/5

5. There are 2 9/12 of the pies left.

6. a. 1 1/5   b. 1 2/5   c. 2/5

7. a. 1 2/4   b. 2 6/8   c. 3 2/6
   d. 4/5   e. 4 3/5   f. 2 2/3

8. It is 1 inch. Half of the perimeter is 2 ½ inches, and the two sides add up to that (1 in + 1 ½ in = 2 ½ in).

9.

a. $\frac{6}{10} - \frac{15}{100}$ $\downarrow \quad \downarrow$ $\frac{60}{100} - \frac{15}{100} = \frac{45}{100}$	b. $\frac{7}{10} - \frac{38}{100}$ $\downarrow \quad \downarrow$ $\frac{70}{100} - \frac{38}{100} = \frac{32}{100}$	c. $\frac{54}{100} - \frac{2}{10}$ $\downarrow \quad \downarrow$ $\frac{54}{100} - \frac{20}{100} = \frac{34}{100}$

10. 6 − 2 2/3 = 3 1/3. The part left is 3 1/3 yards long. In feet, 2 2/3 yards = 8 feet, and 3 1/3 feet = 10 feet.

11. 11/12 of a pizza is left. Edward, Abigail, Jack, and John ate 13 pieces, which is 1 1/12 of a pizza. Since Mom and Dad ate 1 pizza, in total 2 1/12 of the pizzas were consumed. So, 11/12 of a pizza is left.

12. a. 6 − 2 2/3 = 3 1/3. There is 3 1/3 cups of flour left.
   b. She can make one more batch.

Puzzle corner. Half of the perimeter is 3 ¼ inches. We can write the addition: 1 ¾ + (?) = 3 ¼. Solution: (?) = 1 2/4 or 1 ½. The other side is 1 ½ inches.

Comparing Fractions, p. 161

1. a. <   b. >   c. <   d. >

2. a. >   b. >   c. >   d. <

3. a. <   b. >   c. >   d. >   e. >   f. <   g. =   h. <

4. a. $\dfrac{3}{8}$, $\dfrac{3}{6}$, $\dfrac{6}{8}$ b. $\dfrac{2}{5}$, $\dfrac{5}{6}$, $\dfrac{6}{5}$ c. $\dfrac{1}{7}$, $\dfrac{1}{4}$, $\dfrac{5}{8}$

5. a. >   b. <   c. <   d. <

6. a. <   b. <   c. <   d. <   e. >   f. <   g. >   h. >   i. >   j. >

7.

a.	$\dfrac{1}{5}\quad\dfrac{3}{10}$ $\downarrow\qquad\downarrow$ $\dfrac{2}{10} < \dfrac{3}{10}$	b.	$\dfrac{3}{4}\quad\dfrac{5}{8}$ $\downarrow\qquad\downarrow$ $\dfrac{6}{8} > \dfrac{5}{8}$	c.	$\dfrac{5}{12}\quad\dfrac{1}{3}$ $\downarrow\qquad\downarrow$ $\dfrac{5}{12} > \dfrac{4}{12}$	d.	$\dfrac{11}{12}\quad\dfrac{5}{6}$ $\downarrow\qquad\downarrow$ $\dfrac{11}{12} > \dfrac{10}{12}$
e.	$\dfrac{3}{4}\quad\dfrac{9}{12}$ $\downarrow\qquad\downarrow$ $\dfrac{9}{12} = \dfrac{9}{12}$	f.	$\dfrac{5}{9}\quad\dfrac{2}{3}$ $\downarrow\qquad\downarrow$ $\dfrac{5}{9} < \dfrac{6}{9}$	g.	$\dfrac{1}{3}\quad\dfrac{2}{9}$ $\downarrow\qquad\downarrow$ $\dfrac{3}{9} > \dfrac{2}{9}$	h.	$\dfrac{3}{12}\quad\dfrac{1}{3}$ $\downarrow\qquad\downarrow$ $\dfrac{3}{12} < \dfrac{4}{12}$

8. a. cannot compare   b. 3/9 = 2/6   c. 7/10 > 5/8   d. cannot compare   e. cannot compare   f. cannot compare

9. $\dfrac{1}{3}$, $\dfrac{3}{8}$, $\dfrac{2}{5}$, $\dfrac{5}{8}$, $\dfrac{2}{3}$

10. Answers will vary. The student can use number lines, bars, circles, or other shapes. For example:

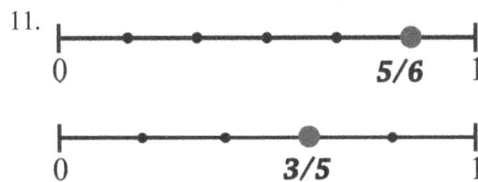

11.

12. Angie ate more pizza. She ate 1/8 of the pizza more than Joe. That is because Joe ate 1/4 = 2/8 of the pizza.

13. Chloe does. She pays 3/10, which is 30/100, of her paycheck in taxes.

14. If it is discounted 4/10 of its price, because 4/10 = 40/100.

15. a. The wholes are not the same size.

   b. The student can use number lines, bars, circles, or other shapes. Using pie pictures, we get

16.

a. $\dfrac{3}{7}$, $\dfrac{3}{5}$, $1\dfrac{1}{7}$	b. $\dfrac{3}{8}$, $\dfrac{3}{6}$, $1\dfrac{1}{4}$	c. $\dfrac{4}{9}$, $\dfrac{2}{3}$, $\dfrac{6}{5}$

Puzzle Corner. Dad ate more. Eating 2/3 of the smaller pizza, which is 1/2 the size of the larger pizza, is equal to eating 1/3 of the larger pizza. Dad ate 3/8 of the larger pizza. Now, 3/8 > 1/3 (see exercise #9), so Dad ate more pizza.

1.

a. $\dfrac{3}{7} = 3 \times \dfrac{1}{7}$	b. $\dfrac{6}{9} = 6 \times \dfrac{1}{9}$	c. $4 \times \dfrac{1}{5} = \dfrac{4}{5}$	d. $7 \times \dfrac{1}{10} = \dfrac{7}{10}$

2.

a. $\dfrac{8}{7} = 8 \times \dfrac{1}{7}$	b. $1\dfrac{3}{5} = \dfrac{8}{5} = 8 \times \dfrac{1}{5}$	c. $1\dfrac{2}{3} = \dfrac{5}{3} = 5 \times \dfrac{1}{3}$
d. $10 \times \dfrac{1}{6} = \dfrac{10}{6} = 1\dfrac{4}{6}$	e. $7 \times \dfrac{1}{4} = \dfrac{7}{4} = 1\dfrac{3}{4}$	f. $9 \times \dfrac{1}{3} = \dfrac{9}{3} = 3$

3. a. $10 \times 1/3 = 10/3 = 3\ 1/3$. She needs to buy at least 3 1/3 lb of chicken.
   b. Between 3 and 4.
   c. $10 \times 1/2 = 5$. She needs 5 quarts of juice.

4.

a. $3 \times \dfrac{2}{4} = \dfrac{6}{4} = 1\dfrac{2}{4}$	b. $4 \times \dfrac{2}{6} = \dfrac{8}{6} = 1\dfrac{2}{6}$	c. $2 \times \dfrac{7}{8} = \dfrac{14}{8} = 1\dfrac{6}{8}$

5.

a. $5 \times \dfrac{3}{8} = \dfrac{15}{8} = 1\dfrac{7}{8}$	b. $4 \times \dfrac{2}{5} = \dfrac{8}{5} = 1\dfrac{3}{5}$
c. $5 \times \dfrac{7}{12} = \dfrac{35}{12} = 2\dfrac{11}{12}$	d. $5 \times \dfrac{6}{10} = \dfrac{30}{10} = 3$
e. $9 \times \dfrac{5}{8} = \dfrac{45}{8} = 5\dfrac{5}{8}$	

Can you find a shortcut for these problems?
Answers will vary. For example: Multiply the top number of the fraction by the whole number.

f. $4 \times \dfrac{2}{3} = \dfrac{8}{3} = 2\dfrac{2}{3}$	g. $3 \times \dfrac{4}{10} = \dfrac{12}{10} = 1\dfrac{2}{10}$	h. $2 \times \dfrac{5}{6} = \dfrac{10}{6} = 1\dfrac{4}{6}$

6.

a. $\dfrac{8}{5} = 4 \times \dfrac{2}{5}$	b. $\dfrac{9}{4} = 3 \times \dfrac{3}{4}$	c. $2\dfrac{2}{3} = 2 \times 1\dfrac{1}{3}$

7. a. 1 1/4   b. 2   c. 1 1/7   d. 1 4/10   e. 2 2/8   f. 14/100
   g. 2 1/10   h. 72/100   i. 3 3/10   j. 3 1/8   k. 2 2/3   l. 3 2/4

8. $4 \times 7/8$ in = 28/8 in = 3 4/8 in (which also equals 3 1/2 in).

9. a. $5 \times 1\ 1/8$ in = 5 5/8 in.
   b. Double the previous result to get 10 10/8 in = 11 2/8 in.

10. Meat: $8 \times 1/4$ lb = 2 lb. Pasta: $8 \times 3/4$ C = 24/4 C = 6 C.

1.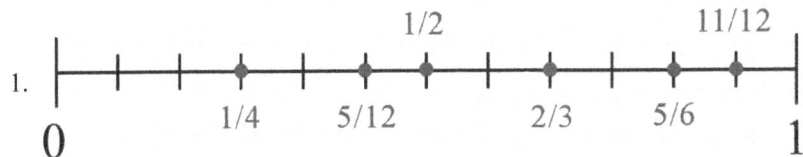

2.

a. $\frac{2}{6}, \frac{1}{2}, \frac{2}{3}$	b. $\frac{1}{8}, \frac{1}{4}, \frac{3}{8}$
c. $\frac{2}{5}, \frac{1}{2}, \frac{3}{5}$	d. $\frac{3}{8}, \frac{3}{4}, \frac{4}{5}$

3. $3 \times 3/5$ mi = $9/5$ mi = $1\ 4/5$ mi.

4. Answers may vary.

a. $\frac{1}{3} = \frac{2}{6}$ $\frac{4}{12}$		b. $\frac{1}{4} = \frac{2}{8}$ $\frac{3}{12}$	
c. $\frac{3}{4} = \frac{6}{8}$ $\frac{9}{12}$		d. $\frac{2}{5} = \frac{4}{10}$ $\frac{6}{15}$	

5. a. 4/7   b. 3 1/3   c. 2 1/4   d. 2   e. 7   f. 2

6. a. 1   b. 1 3/8   c. 7 1/5   d. 8/12   e. 8/10   f. 4 2/4

7. Multiplying by one-half is actually the same as _dividing_ by 2.

a.	b.	c.
$2 \times \frac{1}{2} = 1$	$7 \times \frac{1}{2} = 3\frac{1}{2}$	$15 \times \frac{1}{2} = 7\frac{1}{2}$
$3 \times \frac{1}{2} = 1\frac{1}{2}$	$8 \times \frac{1}{2} = 4$	$20 \times \frac{1}{2} = 10$
$4 \times \frac{1}{2} = 2$	$9 \times \frac{1}{2} = 4\frac{1}{2}$	$17 \times \frac{1}{2} = 8\frac{1}{2}$
$5 \times \frac{1}{2} = 2\frac{1}{2}$	$10 \times \frac{1}{2} = 5$	$21 \times \frac{1}{2} = 10\frac{1}{2}$
$6 \times \frac{1}{2} = 3$	$11 \times \frac{1}{2} = 5\frac{1}{2}$	$32 \times \frac{1}{2} = 16$

Puzzle Corner.
a. 1/2 + 3/8 = 4/8 + 3/8 = 7/8
b. 1/3 + 1/6 = 2/6 + 1/6 = 3/6
c. 1/3 + 2/9 = 3/9 + 2/9 = 5/9

Mixed Review Chapter 7, p. 170

1.

a. $57 \div 5 = 11$ R2	b. $34 \div 7 = 4$ R6	c. $33 \div 9 = 3$ R6
$11 \times 5 + 2 = 57$	$4 \times 7 + 6 = 34$	$3 \times 9 + 6 = 33$

2. a. 7 in $\times$ 3 in $= 21$ in^2 b. 25 km $\times$ 20 km $= 500$ km^2 c. 2 ft $\times$ 9 1/2 ft $= 19$ ft^2

3. a. acute   b. right   c. obtuse   d. right   e. obtuse   f. acute   g. right   h. right

4. a. Answers will vary. Check the student's parallelogram. For example:
   b. Answers will vary. The measurements in this picture are not to scale.
   c. Answers will vary. The angles of the parallelogram on the right are
      58°, 122°, 58°, and 122°.

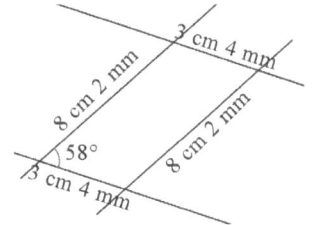

5. a. 268   b. 277

6. a. 4×44 lb $= 176$ lb and 5×32 lb $= 160$ lb.
      <u>Four boxes 44 lb each</u> weigh more. They weigh 16 lb more than five boxes 32 lb each.
   b. $86 \div 25 = 3$ R11. The teacher got to keep <u>11 balloons</u>.
   c. $\$9.73 \div 7 = \$1.39$ and $5 \times \$1.39 = \$6.95$. Five liters would cost <u>$6.95</u>.

7. $\$20 - (\$7 + \$5) = \$8$.

Review Chapter 7, p. 172

1. a. 1   b. 5 1/8   c. 7   d. 2/10   e. 1 2/4   f. 6 7/12

2. a. 7/10   b. 3/5   c. 4/5   d. 5/8

3. a. 33/100   b. 53/100   c. 1 17/100

4. 1 3/4 liters

5. Answers will vary. For example: ⬡ = ⬡

6.

a.	b.	c. $\dfrac{2}{5} = \dfrac{4}{10}$	d. $\dfrac{2}{3} = \dfrac{6}{9}$
$\times 2$ $\dfrac{3}{4} = \dfrac{6}{8}$ $\times 2$	$\times 5$ $\dfrac{1}{2} = \dfrac{5}{10}$ $\times 5$	e. $\dfrac{2}{3} = \dfrac{8}{12}$	f. $\dfrac{3}{4} = \dfrac{12}{16}$

7. a. >   b. <   c. =   d. >   e. >   f. >   g. >   h. <

8. a. 9/10   b. 1 1/5   c. 1 4/10   d. 99/100   e. 2 4/8   f. 2 9/12

9.

> **Mexican Coffee (4x)**
>
> 6 cups strong gourmet coffee
> 3 tsp cinnamon
> 16 tsp chocolate syrup
> 1 tsp nutmeg
> 2 cup heavy cream
> 4 tbsp sugar

10. a. 40; 80   b. 8 cm; 40 cm   c. 400 kg; $40

11. Since he has 1/4 of his birthday money left, he has <u>$5</u> left.

12. One-eighth of 240 pages is 30 pages. She has 5/8 of the book left to read, which is $5 \times 30 = $ <u>150 pages</u>.

Chapter 8: Decimals

Decimal Numbers—Tenths, p. 177

1. a. 0.7   b. 2.4   c. 10.9   d. 9/10   e. 29 3/10

2. a. 0.6 = 6/10   b. 1.2 = 1 2/10   c. 2.9 = 2 9/10

3.

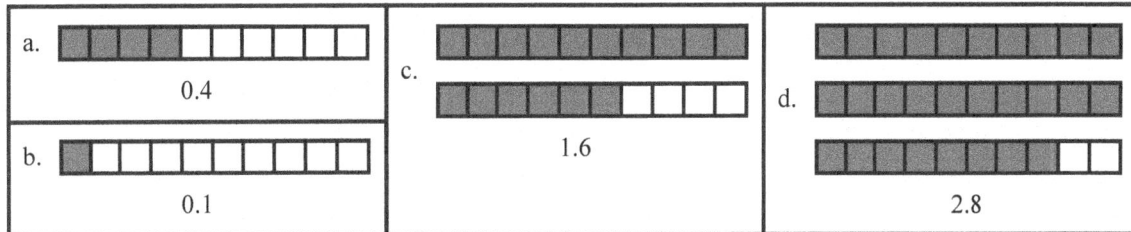

4. 7, 7 1/10, 7 2/10, 7 3/10, 7 4/10, 7 5/10, 7 6/10, 7 7/10, 7 8/10, 7 9/10, 8

5. a.

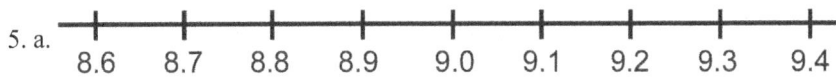

b.

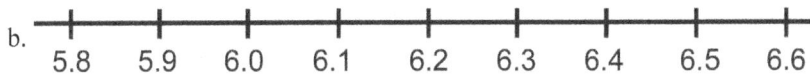

6. a.

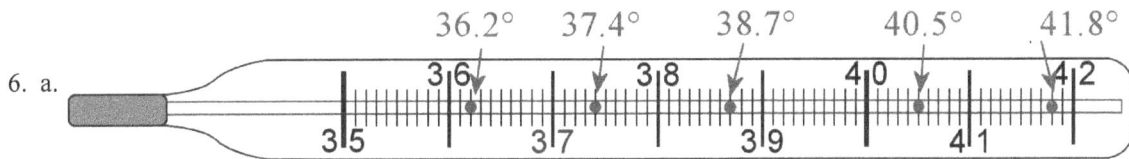

   b. The temperatures 38.7°, 40.5°, and 41.8° are fever.

7. a. <   b. >   c. >   d. <   e. =

8. 0.1 $\frac{1}{2}$ 0.9   1.2   2.3 $2\frac{1}{2}$ 2.6   3.0

Adding and Subtracting with Tenths, p. 179

1. a. 0.7 + 0.5 = 1.2
   b. 0.6 + 0.8 = 1.4
   c. 1.1 − 0.8 = 0.3
   d. 1.3 − 0.4 = 0.9
   e. 0.2 + 1.1 = 1.3

2. a. 9/10;   0.2 + 0.7 = 0.9
   b. 1 1/10;   0.5 + 0.6 = 1.1
   c. 1 7/10;   0.9 + 0.8 = 1.7

3. a. 1.1; 2.1
   b. 1.2; 4.2
   c. 1.5; 3.5
   d. 0.9; 4.9

4. a. 3.2   b. 2.2   c. 6.1   d. 3

5. a. 5.3   b. 78.4   c. 63.4

6.

a. 0.1	b. 1.1	c. 2.5	d. 3.6
+ 0.2 = 0.3	+ 0.5 = 1.6	+ 0.3 = 2.8	− 0.4 = 3.2
+ 0.2 = 0.5	+ 0.5 = 2.1	+ 0.3 = 3.1	− 0.4 = 2.8
+ 0.2 = 0.7	+ 0.5 = 2.6	+ 0.3 = 3.4	− 0.4 = 2.4
+ 0.2 = 0.9	+ 0.5 = 3.1	+ 0.3 = 3.7	− 0.4 = 2
+ 0.2 = 1.1	+ 0.5 = 3.6	+ 0.3 = 4.0	− 0.4 = 1.6
+ 0.2 = 1.3	+ 0.5 = 4.1	+ 0.3 = 4.3	− 0.4 = 1.2

7. a.

  b. 2.4 cm

8. a. 5 mm; 12 mm
   b. 0.7 cm; 3.5 cm
   c. 1.4 cm; 7.4 cm

9. See the rectangle on the right. The perimeter is 20.2 cm.

Two Decimal Digits—Hundredths, p. 181

1. a. $0.08 = 8/100$ b. $0.55 = 55/100$ c. $1.50 = 1\ 50/100$ d. $1.06 = 1\ 6/100$ e. $3.70 = 3\ 70/100$

2.

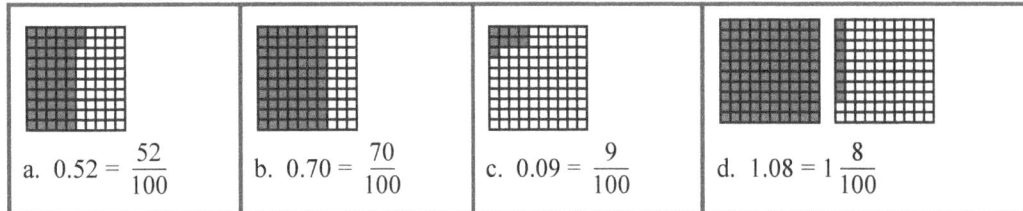

a. $0.52 = \dfrac{52}{100}$ b. $0.70 = \dfrac{70}{100}$ c. $0.09 = \dfrac{9}{100}$ d. $1.08 = 1\dfrac{8}{100}$

Teaching box:

Now, *draw* nine tiny lines between 0.2 and 0.3, dividing that distance into TEN new parts.

If this process was repeated between 0.3 and 0.4, between 0.4 and 0.5, and so on, into
<u>how many parts</u> in total would the number line from 0 to 1 be divided?   <u>100</u> parts

These new parts are therefore **hundredth parts**, or **hundredths**.

3.

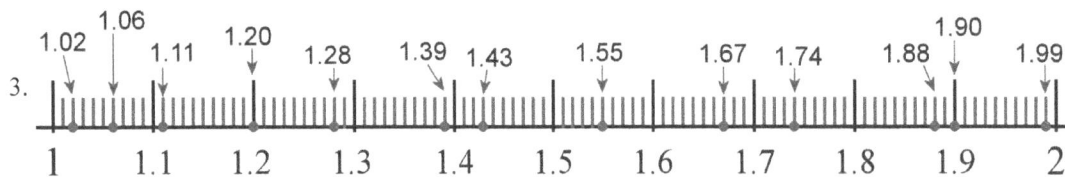

4. 3.60, 3.61, 3.62, 3.63, 3.64, 3.65, 3.66, 3.67, 3.68, 3.69, 3.70

5. a. 0.01; 0.1
   b. 0.04; 0.4
   c. 0.31; 0.3
   d. 2.03; 2.3
   e. 7.5; 5.17
   f. 10.1, 10.01

98

Two Decimal Digits—Hundredths, continued

6.

	fraction	read as ...
a. 0.02	2/100	two hundredths
b. 1.49	1 49/100	one and forty-nine hundredths
c. 5.5	5 5/10	five and five tenths
d. 3.08	3 8/100	three and eight hundredths

7.

a. $0.50 = 0.5$

$\dfrac{50}{100} = \dfrac{5}{10}$

b. $0.10 = \underline{0.1}$

$\dfrac{10}{100} = \dfrac{1}{10}$

c. $0.80 = \underline{0.8}$

$\dfrac{80}{100} = \dfrac{8}{10}$

8. $0.60

9. a. >   b. <   c. <   d. =

10. a. 7.9   b. 15.4 and 15.40 (they are equal)   c. 2.77   d. 9.3   e. 3.6   f. 0.4

11. a. >   b. =   c. >   d. =   e. <   f. >   g. <   h. >   i. >   j. <   k. >   l. <

12. a. 5.06 < 5.16 < 5.6 < 5.66
   b. 7.70 < 7.77 < 7.78 < 7.8

Add and Subtract Decimals in Columns, p. 185

1. a. 62.29   b. 19.28   c. 183.39

2. a. 13.99   b. 49.89   c. 12.16

3. a. >   b. <   c. <   d. <   e. >   f. <   g. <   h. >

4. a. 14.03   b. 70.64   c. 4.84

5. a. 0.82   b. 15.63

6.

a. Mary did not line up the decimal points correctly. The correct answer:	b. Jack did not regroup at all. The correct answer:
$\begin{array}{r} 4\ 5\ .\ 5 \\ +\ \ \ 5\ .\ 3\ 4 \\ \hline 5\ 0\ .\ 8\ 4 \end{array}$	$\begin{array}{r} 9 \quad\ \ 9 \\ 8\ \cancel{10}\ \ \cancel{10}\ 10 \\ \cancel{9}\ \cancel{0}\ .\ \cancel{0}\ \cancel{0} \\ -\ 8\ 8\ .\ 5\ 6 \\ \hline 1\ .\ 4\ 4 \end{array}$

7. 2.78 kg

8. 11.25 m

9. 2.65 kg

Puzzle corner   a. 4.8 + 40.8 + 4.08 = 49.68   b. 560 − 5.06 − 56 = 498.94

1.

a. $0.05 + 0.04 = 0.09$ $\dfrac{5}{100} + \dfrac{4}{100} = \dfrac{9}{100}$	b. $0.07 + 0.04 = 0.11$ $\dfrac{7}{100} + \dfrac{4}{100} = \dfrac{11}{100}$	c. $0.37 - 0.06 = 0.31$ $\dfrac{37}{100} - \dfrac{6}{100} = \dfrac{31}{100}$
d. $0.45 + 0.65 = 1.10$ $\dfrac{45}{100} + \dfrac{65}{100} = 1\dfrac{10}{100}$	e. $3.25 - 1.08 = 2.17$ $3\dfrac{25}{100} - 1\dfrac{8}{100} = 2\dfrac{17}{100}$	

2. a. 0.12; 4.12
   b. 0.95; 2.80
   c. 1; 2.02

3.

a. 0.80	b. 2.90	c. 1.77
$- 0.05 = 0.75$	$+ 0.03 = 2.93$	$+ 0.11 = 1.88$
$- 0.05 = 0.70$	$+ 0.03 = 2.96$	$+ 0.11 = 1.99$
$- 0.05 = 0.65$	$+ 0.03 = 2.99$	$+ 0.11 = 2.10$
$- 0.05 = 0.60$	$+ 0.03 = 3.02$	$+ 0.11 = 2.21$
$- 0.05 = 0.55$	$+ 0.03 = 3.05$	$+ 0.11 = 2.32$
$- 0.05 = 0.50$	$+ 0.03 = 3.08$	$+ 0.11 = 2.43$

4. a. 0.96   b. 0.87   c. 0.63
   d. 2.95   e. 4.92   f. 8.46

5.

a. $0.97 + 0.08 = 1.05$ $\dfrac{97}{100} + \dfrac{8}{100} = \dfrac{105}{100}$	b. $0.92 + 0.09 = 1.01$ $\dfrac{92}{100} + \dfrac{9}{100} = \dfrac{101}{100}$	c. $0.91 + 0.12 = 1.03$ $\dfrac{91}{100} + \dfrac{12}{100} = \dfrac{103}{100}$
d. $1.03 - 0.04 = 0.99$ $\dfrac{103}{100} - \dfrac{4}{100} = \dfrac{99}{100}$	e. $1.12 - 0.16 = 0.96$ $\dfrac{112}{100} - \dfrac{16}{100} = \dfrac{96}{100}$	f. $1.06 - 0.09 = 0.97$ $\dfrac{106}{100} - \dfrac{9}{100} = \dfrac{97}{100}$

6. a. 1.01   b. 3.02   c. 0.96
   d. 1.02   e. 4.06   f. 0.97

Teaching box:

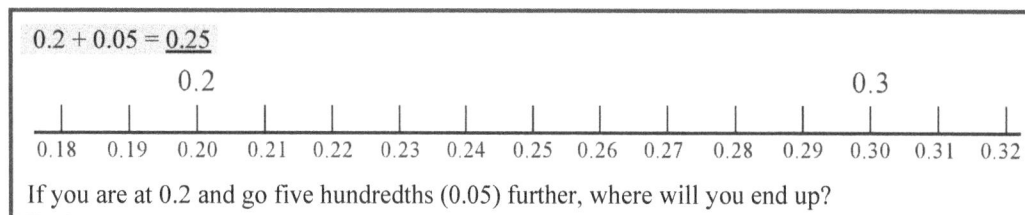

$0.2 + 0.05 = \underline{0.25}$

If you are at 0.2 and go five hundredths (0.05) further, where will you end up?

7.

a. $0.7 + 0.04$	b. $0.5 + 0.11$
$\downarrow \quad\quad \downarrow$	$\downarrow \quad\quad \downarrow$
$0.70 + 0.04 = 0.74$	$0.50 + 0.11 = 0.61$

8.

a. $0.1\underline{0} + 0.05 = \underline{0.15}$ $\dfrac{10}{100} + \dfrac{5}{100} = \dfrac{15}{100}$	b. $0.04 + 0.4\underline{0} = 0.44$ $\dfrac{4}{100} + \dfrac{40}{100} = \dfrac{44}{100}$	c. $0.6\underline{0} - 0.09 = 0.51$ $\dfrac{60}{100} - \dfrac{9}{100} = \dfrac{51}{100}$
d. $0.6\underline{0} + 0.22 = \underline{0.82}$ $\dfrac{60}{100} + \dfrac{22}{100} = \dfrac{82}{100}$	e. $0.73 - 0.5\underline{0} = 0.23$ $\dfrac{73}{100} - \dfrac{50}{100} = \dfrac{23}{100}$	f. $0.9\underline{0} - 0.13 = 0.77$ $\dfrac{90}{100} - \dfrac{13}{100} = \dfrac{77}{100}$

9.

a. $0.11 + 0.5\underline{0} = 0.61$	b. $0.24 - 0.2\underline{0} = 0.04$	c. $0.3\underline{0} + 0.39 = 0.69$
d. $0.22 + 0.7\underline{0} = 0.92$	e. $0.6\underline{0} - 0.41 = 0.19$	f. $0.97 - 0.7\underline{0} = 0.27$

10. It overswept.

11.

1.6	1.21	1.3	1.3	1.18	1.45	1.48
1.52	1.17	1.24	0.7	1.24	1.37	1.4
1.44	1.14	1.18	1.12	1.6	1.31	1.34
1.23	0.94	1.02	1.06	0.98	1.25	1.28
1.2	0.9	0.94	1	0.92	1.21	1.22
0.76	0.82	0.88	0.4	0.86	1.16	1.18
0.7	0.74	0.84	1.71	0.8	1.08	1.12

Puzzle corner. a. $x = 0.15$ b. $x = 0.06$ c. $x = 0.18$

Using Decimals with Measuring Units, p. 192

1. a. $400 \text{ m} = 0.4 \text{ km}$ b. $700 \text{ m} = 0.7 \text{ km}$

2. a. 0.5 km b. 0.9 km c. 200 m

3. a. 600 m; 1,100 m
   b. 0.7 km; 1.8 km
   c. 10,900 m; 24.6 km

4. a. Amanda walks 0.6 km.

   b. Julie walks $2.4 \text{ km} - 0.6 \text{ km} = \underline{1.8 \text{ km more.}}$

5. Andrew ran a longer distance (2,400 m). He ran $2,400 \text{ m} - 2,040 \text{ m} = 360 \text{ m}$ more.

Using Decimals with Measuring Units, cont.

6. a. 700 ml; 0.7 L
   b. 300 ml; 0.3 L
   c. 200 ml; 500 ml; 5,400 ml
   d. 0.1 L; 1.5 L; 6.3 L

7. a. 0.6 kg; 2.4 kg
   b. 200 g; 800 g
   c. 20.5 kg; 7,100 g

8. There is 700 ml or 0.7 liters of juice left.

9. There is 7.3 liters, or 7,300 ml, left.

10. a. 350 g
    b. 2,250 g. Change 2.6 kg to 2,600 g, and then subtract 2,600 g – 350 g = 2,250 g.

Mixed Review Chapter 8, p. 194

1. a. 5/8 + 3/8 + 2/8 = 1 2/8
   b. 1 7/12 + 7/12 = 2 2/12

2. a. 50/12 = 4 2/12   b. 28/9 = 3 1/9   c. 42/100

3. a. 1, 2, 19, 38
   b. 1, 2, 4, 7, 8, 14, 28, 56
   c. 1, 19 (it is prime)

4. a. $\approx$ 6 × 300 = 1,800 Exact: 1,752
   b. $\approx$ 11 × 40 = 440   Exact 462
   c. $\approx$ 3 × 2,400 = 7,200   Exact 7092
   d. $\approx$ 7 × 9,000 = 63,000   Exact 61,789

5. a. 90   b. 4   c. 30

6. a. 2:30 pm   b. 7:15 pm   c. 10:45 pm   d. 7:50 am

7. Mom paid $24.00 and Terry paid $72.00.

8. First find 1/5 of $600, which is $120. Jack still has
   to pay 2/5 of the price, which is 2 × $120 = $240.

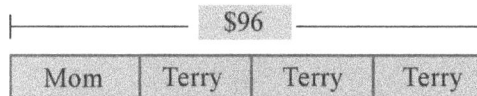

9.

10. a.   b. 97°

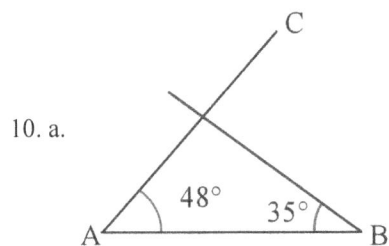

102

1. a. 0.7   b. 0.07   c. 1.6   d. 2.41   e. 1.01   f. 0.47   g. 8/10   h. 2 9/10   i. 4 14/100   j. 18 8/100   k. 3/100   l. 29/100

2. a. >   b. >   c. <   d. =   e. >   f. =   g. <   h. <   i. >

3.

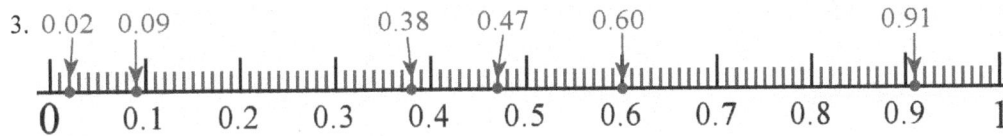

4. 0.1, 0.12, 0.2, 0.21, 1/2, 0.74, 0.8

5.

a. 0.7<u>0</u> + 0.03 = 0.73	b. 0.32 + 0.4<u>0</u> = 0.72	c. 0.7<u>0</u> − 0.04 = 0.66
$\dfrac{70}{100} + \dfrac{3}{100} = \dfrac{73}{100}$	$\dfrac{32}{100} + \dfrac{40}{100} = \dfrac{72}{100}$	$\dfrac{70}{100} - \dfrac{4}{100} = \dfrac{66}{100}$

6. a. 1   b. 0.88   c. 0.36   d. 0.24   e. 0.83   f. 0.5

7. a. Incorrect. Should be: 0.99 + 0.1 = 1.09   OR   0.99 + 0.01 = 1.
   b. Correct.
   c. Incorrect. Should be: 0.19 + 0.19 = 0.38.
   d. Incorrect. Should be: 0.03 + 0.5 = 0.53   OR   0.03 + 0.05 = 0.08.

8. a. 9.31   b. 23.11   c. 5.84

9. 2.84

10. If your temperature is 99.9°F, is 1.3 degrees above the normal body temperature.
    If your temperature is 100.4°F, it is 1.8 degrees above the normal body temperature.

11. The tablet weighing 610 grams is heavier, since 0.6 kg equals 600 grams.

Test Answer Keys

Math Mammoth Grade 4 Tests Answer Key

Chapter 1 Test

1. $x = 2,611$

2. a. 260   b. 20   c. 70

3. The expression $\$20 - 7 \times \$2$ matches the problem. The answer is $6.

4. Estimations may vary because estimation is not an exact "science".
   For example: $\$30 + 2 \times \$14 = \$58$ (rounding the first number up to $30 for easier mental adding,
   rounding the other down to the nearest dollar) OR
   $\$29 + 2 \times \$14 = \$57$ (rounding everything to the nearest dollar).

5. 10 m + 8 m + 15 m = 33 m

6. $\$67 + \$48 = x$; $x = \$115$.

Chapter 2 Test

1. a. 400,040   b. 64,500   c. 200,067

2. a. eighty thousand or 80,000   b. eighty or 80

3. a. 516,800   b. 293,000   c. 200,000

4. 207,698

5. 3,294   39,244   39,294   93,294   399,295

6. The king of Nootyland has more coins; he has 5,218 coins more. The king of Sookiland has $3 \times 24,000 + 1,382$
   $= 73,382$ coins. The king of Nootyland has 78,600 coins; the difference is $78,600 - 73,382 = 5,218$.

7. A thousand students.

Chapter 3 Test

1. a. $40 + 32 = 72$
   b. $140 + 42 = 182$
   c. $2,100 + 27 = 2,127$

2. About $7 \times \$20 = \$140$.

3. a. 6,300   b. 6,000   c. 160,000

4. a. 200   b. 200   c. 8

5. a. 18,000   b. 48,000   c. 1,293   d. 2,080

6. a. 1,170   b. 5,848   c. 1,045   d. 15,924

7. 1,360

8. a. One meal costs $3, so seven meals cost $21.
   b. $\$30 - 7 \times \$2.55 = \$30 - \$17.85 = \$12.15$.
   c. She took $3 \times \$12.55 + \$8.90 + \$13.45 = \$60$.
   d. It would have cost $30 ($150 ÷ 5 = 30).

Chapter 4 Test

1. 3:50 p.m.

2. a. 2 3/8 in. or 6 cm 0 mm   b. 3 7/8 in. or 9 cm 8 mm

3.

a.	b.	c.
4 lb 2 oz = 66 oz 76 cm = 760 mm 5 ft 5 in = 65 in.	2 L 80 ml = 2,080 ml 3 qt = 12 cups 200 yd = 600 ft	7 m 5 cm = 705 cm 4 kg 500 g = 4,500 g 3 T = 6,000 lb

4. 10 cm 4 mm

5. a. two bottles
   b. five bottles.

6. a. $7.60
   b. $10.50

7. eight jars

Chapter 5 Test

1. a. 3 R1; 2 R3
   b. 4 R5; 5 R5
   c. 3 R4; 7 R4

2. One meter costs $6, so five meters would cost 5 × $6 = $30.

3. $210. One-fifth of $350 is $70; three-fifths of that is 3 × $70 = $210.

4. 250 bricks. One-third of his 1,200 bricks is 400 bricks, and two-thirds of them is double that, or 800 bricks. So, he has 400 bricks left. After selling 150 bricks, he has 250 bricks left.

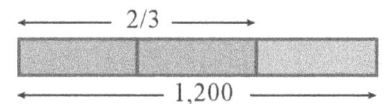

5. a. 113. Check: 5 × 113 = 565   b. 458 Check: 8 × 458 = 3,664

6. a. 1, 2, 4, 7, 14, 28
   b. 1, 13
   c. 1, 2, 4, 8, 16, 32
   d. 1, 2, 4, 19, 38, 76

7. 125 ÷ 7 = 17 R6. Each child got 17 pencils, and 6 pencils were left over.

8. $47. Add the prices and divide by four: average = ($39 + $45 + $63 + $41) ÷ 4 = $47.

9. Yes, it is, because 924 ÷ 7 = 132 and there is no remainder; the division is even.

10. (30 − 10) × 20 = 400

Chapter 6 Test

1. Check the student's answer. Here is a 75° angle:

2. a. 140°

3. It is 148 degrees.

   $78° + 134° + x = 360°$

   $x = 360° − 134° − 78°$

   $= 148°$

4. a. a parallelogram or a rhombus (either is correct)

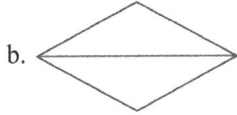

b.

c. 17 cm 7 mm

5. Answers will vary. Check the student's answer. For example:

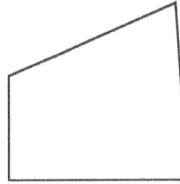

6. Answers will vary. Check the student's answers. The two other angles should have an angle sum of 90°.

7. a. a right triangle
   b. an acute triangle

8. Subtract the area of the outer rectangle and the area of the white rectangle: 20 ft × 10 ft − 10 ft × 3 ft
   = 200 ft^2 − 30 ft^2 = 170 ft^2.

Chapter 7 Test

1. a. 1   b. 2 1/3   c. 5 4/5   d. 2/12   e. 2 3/5   f. 5 4/6

2. a. $\dfrac{3}{8}, \dfrac{1}{2}, \dfrac{3}{4}$ b. $\dfrac{5}{7}, \dfrac{5}{5}, \dfrac{7}{5}$ c. $\dfrac{5}{9}, \dfrac{5}{6}, \dfrac{5}{2}$

3.

a. Split all pieces into four new ones.	b. Split all pieces into three new ones.
$\dfrac{1}{2} = \dfrac{4}{8}$	$\dfrac{2}{3} = \dfrac{6}{9}$

4.

a. $\dfrac{1}{5} = \dfrac{2}{10}$	b. $\dfrac{3}{4} = \dfrac{9}{12}$	c. $\dfrac{4}{5} = \dfrac{20}{25}$	d. $\dfrac{1}{6} = \dfrac{4}{24}$

5. a. 12/10 = 1 2/10
   b. 15/5 = 3
   c. 12/8 = 1 4/8 (which is equal to 1 1/2)

6. a. Walter and Eric ate equal amounts. Walter ate 1/4 of it, which is equal to 3/12, or 3 pieces.
   b. Eric ate 3/12 and John ate 1/12. So, Eric ate 2/12 of the pizza more than John.

Chapter 8 Test

1.

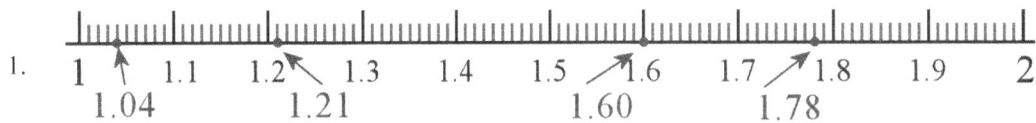

2. a. 0.2   b. 7.04   c. 0.74   d. 52/100   e. 3 9/10

3. a. 2.2   b. 0.95   c. 0.19   d. 0.7   e. 0.37   f. 3.04

4. a. >   b. =   c. >   d. <   e. <

5.   2.07 < 2.17 < 2.7 < 2.77 < 7.2

6. 5.2 kg. You can add 1.3 kg + 1.3 kg + 1.3 kg + 1.3 kg = 5.2 kg

7. a. 7.36   b. 1.76

Grade 4 End-of-the-Year Test Answer Key

1. 1,980. Add to check: 1,980 + 543 + 2,677 = 5,200.

2. a. ≈ $1 + $9 + $4 + $9 = $23
   b. Her bill is $1.28 + $8.92 + $3.77 + $9.34 = $23.31. Her change is $30 − $23.31 = $6.69.

3. Estimate: 5 × $0.90 + 2 × $1.20 = $4.50 + $2.40 = $6.90.

4. a. 30; 84   b. 11; 14   c. 140; 19

5. a. $35 + x = $92 ; x = $57
   b. x − 24 = 37 ; x = 61

6. a. 2,000   1,750   1,500   1,250   1,000   750   500   250

   b. 200, 500, 800, 1100, 1400, 1700

7. In the frequency table we list how many students got that score.

Quiz score	Frequency
1	0
2	1
3	0
4	1
5	3
6	5
7	5
8	4
9	3
10	2

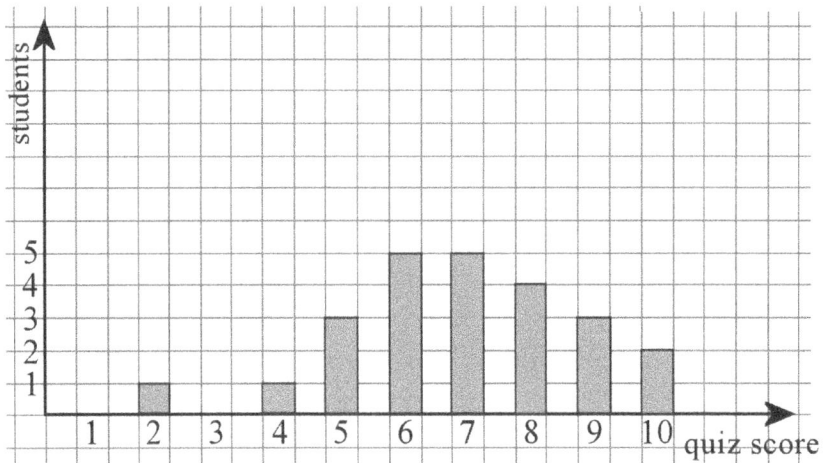

8.

Rubber boots used to cost $27.95 but now the price is $21.45. How much is the discount?

$21.45 + x = $27.95   OR x = $27.95 − $21.45

x = $6.50

←— original price $27.95 —→

| $21.45 | x |

9. a. 1,999   b. 4,980   c. 5,700

110

10. a. 800,050   b. 25,407

11. a. 30,000   b. 9,000   c. 600

12. a. <   b. >   c. >

13. 27,200   217,200   227,200   227,712

14. a. 440,000   b. 90,000   c. 27,500

15. a. 430,000   b. 500,000   c. 10,000

16. a. 501,663   b. 323,688

17. a. 210   b. 4,800   c. 3,200   d. 120   e. 80   f. 70

18. a. $160
    b. $800
    c. four days, since 4 × $160 = $640

19. a. Estimate 5 × 200 = 1,000. Exact: 980
    b. Estimate 40 × 40 = 1,600 or 30 × 40 = 1,200. Exact: 1,330
    c. Estimate 7 × 3,000 = 21,000. Exact: 22,316
    d. Estimate 90 × 20 = 1,800. Exact: 1,958

20.

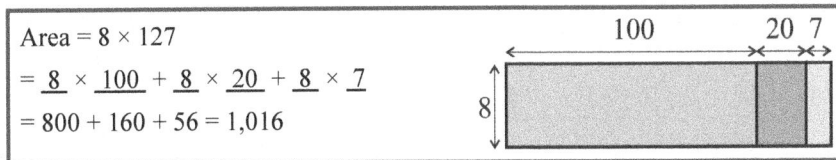

Area = 8 × 127				
= _8_ × _100_ + _8_ × _20_ + _8_ × _7_				
= 800 + 160 + 56 = 1,016				

21. a. Answers will vary. For example: $400 − 26 × $14 = $400 − $364 = $36. Or, 26 × $14 = $364 and $400 − $364 = $36.
    b. 24 × 60 minutes = 1,440 minutes.
    c. Answers will vary. For example: 4 × 375 cm = 1,500 cm. Or, 375 cm + 375 cm + 375 cm + 375 cm = 1,500 cm.
    d. Answers will vary. For example: ($277 − $58) × 8 = $1,752. Or, $277 − $58 = $219 and 8 × $219 = $1,752.

22. Answers may vary. Please check the student's answer. If you printed the test yourself and didn't print at a scale of 100% (but used "shrink to fit" or "fit to printable area"), the figure is probably slightly smaller than intended.

    a. 5 1/4 in or 13 cm 3 mm. 13 cm 4 mm is also acceptable.
    b. 3 7/8 in or 9 cm 8 mm. 9 cm 9 mm is also acceptable.

23. 6 hours 12 minutes

24. 1 h 45 min + 50 min + 1 h 15 min + 2 h 15 min + 55 min = 4 h 180 min, which is 7 hours.

25. She worked 7 hours 30 minutes. From 7:00 am till 3:35 pm is 8 hours 35 minutes. Subtract from that 65 minutes, or 1 hour 5 minutes, to get 7 hours 30 minutes.

26.

a.	b.	c.
6 lb = 96 oz	5 gal = 20 qt	4 ft 2 in = 50 in
2 lb 11 oz = 43 oz	2 qt = 8 cups	7 yd = 21 ft

27.

a.	b.	c.
2 kg = 2,000 g	5 L 200 ml = 5,200 ml	8 cm 2 mm = 82 mm
11 kg 600 g = 11,600 g	3 m = 300 cm	10 km = 10,000 m

28. In four days he jogs 15 km 200 m.

29. 1 L 650 ml

30. 17 ft 8 in

31. a. 63. Check: 63 × 9 = 567
    b. 2,141. Check: 2141 × 4 = 8,564

32. a. 9 R2   b. 8 R1   c. 6 R3

33. a. Three photos on the last page; five pages were full.
    b. Your neighbor should be $36, because one foot of the fence costs $3.

34. a. It cost $99. First find 1/8 of $264: $264 ÷ 8 = $33. Then to find 3/8 of it, multiply 3 × $33 = $99.
    b. She needs 20 bags. 117 ÷ 6 = 19 R3. Notice she needs a bag also for the three muffins that don't fill a bag.

35.

number	divisible by 1	divisible by 2	divisible by 3	divisible by 4	divisible by 5	divisible by 6	divisible by 7	divisible by 8	divisible by 9	divisible by 10
80	x	x		x	x			x		x
75	x		x		x					
47	x									

36.

a. Is 5 a factor of 60? Yes , because 5 × 12 = 60 .	b. Is 7 a divisor of 43? No , because 43 ÷ 7 = 6 R1 (the division is not even).
c. Is 96 divisible by 4? Yes , because 96 ÷ 4 = 24 (the division is even).	d. Is 34 a multiple of 7? No , because 34 is not in the multiplication table of 7. OR: No, because 34 ÷ 7 = 4 R6; the division is not even. OR: No, because there is no whole number you can multiply by 7 to get 34.

37. Answers will vary. For example: 2, 3, and 5. Here is a list of primes less than 100:
    2 3 5 7 11 13 17 19 23 29 31 37 41 43 47 53 59 61 67 71 73 79 83 89 97

38. a. 1, 2, 4, 7, 8, 14, 28, 56   b. 1, 2, 3, 6, 13, 26, 39, 78

39. 155°

40. Check the student's answer.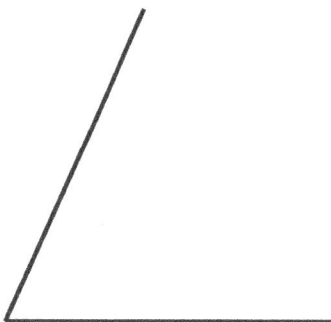

41. Answers will vary. Check the student's answer. The angle sum should be 180° or very close.

42. 29° + x = 180°; x = 151°.

43. Right triangles.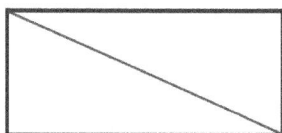

44. Answers will vary. Check the student's answer. For example:

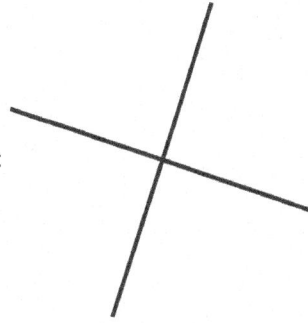

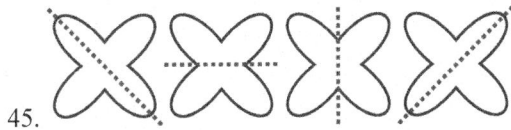

45.

46. Use subtraction: A = 28 ft × 12 ft − 6 ft × 10 ft = 336 ft^2 − 60 ft^2 = 276 ft^2.

47. $\dfrac{5}{8} + \dfrac{5}{8} = 1\dfrac{2}{8}$

48. There is still 2/4 or 1/2 of the puzzle left to do.

49. a. 1 2/5   b. 5/6   c. 6

50.

a. Each piece is split into 2 new ones. $\dfrac{4}{5} = \dfrac{8}{10}$	b. Each piece is split into <u>3</u> new ones. $\dfrac{2}{3} = \dfrac{6}{9}$

51.

a. $\dfrac{2}{3} = \dfrac{10}{15}$	b. $\dfrac{3}{5} = \dfrac{9}{15}$	c. $\dfrac{1}{6} = \dfrac{2}{12}$	d. $\dfrac{1}{3} = \dfrac{3}{9}$

52. a. >   b. >   c. <   d. <

53. $\dfrac{65}{100} < \dfrac{7}{10} < \dfrac{5}{4}$

54. 2 1/4 cups

55. a. 1/8   b. 1 3/5   c. 1 2/12

56.

0.08   0.27   0.55   0.80

0   0.1   0.2   0.3   0.4   0.5   0.6   0.7   0.8   0.9   1

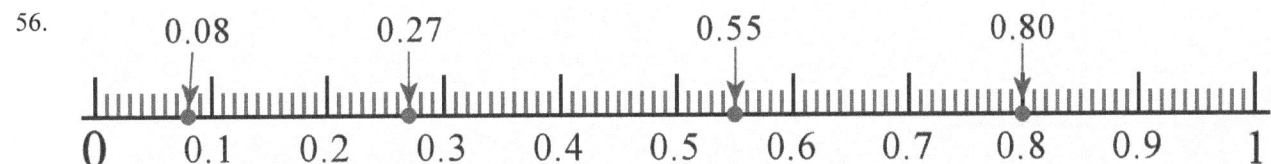

57. a. 0.3   b. 3.9   c. 0.09   d. 7.45

58. a. 6/10   b. 6 7/10   c. 21/100   d. 5 5/100

59. a. <   b. >   c. <   d. =

60. a. 13.01   b. 3.74

Cumulative Reviews
Answer Keys

Cumulative Reviews Answer Key, Grade 4

Cumulative Review: Chapters 1 - 2

1. a. 138; 74; 103   b. 58; 92; 144   c. 127; 70; 144

2. a. 1,800; 1,000   b. 600; 510   c. 9; 1   d. 500; 140

3. a. Continue this pattern: subtract __80__ each time.

700	620	540	460	380	300	220	140

b.

0	99	198	297	396	495	594	693

4. Addition: $450 + 128 + x = 1,000$; Solution $x = 1,000 - 128 - 450 = 422$

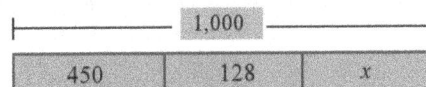

	1,000	
450	128	x

5. a. 4,445   b. 13,378   c. 716,051

6. a. $8,030 < 18,399 < 818,939 < 819,090$ b. $5,220 < 52,200 < 250,500 < 520,500$

7. a. 284 thousand 1   b. 50 thousand 50

2	8	4,	0	0	1

	5	0,	0	5	0

8. a. 2,000   b. 10   c. 500,000   d. 40,000

9. a. 7,500   b. 2,700   c. 4,000   d. 400   e. 56,300   f. 293,600

10. $176 + $25 + $30 = $231.

Cumulative Review: Chapters 1 - 3

1. a. 9; 59; 590; 190   b. 6; 66; 160; 600   c. 5; 75; 500; 550

2. $458 + $366 + $427 + $503 + $413 = $2,167

3. a.

Kilometers	80	400	560	720	800	960	1,200	1,600
Hours	1	5	7	9	10	12	15	20

b.

Dollars	$9	$18	$27	$36	$45	$72	$90	$135
Yards	1	2	3	4	5	8	10	15

4. a. $1,554 < 5,000 < 5,005 < 5,500 < 5,604$ b. $3,800 < 37,700 < 38,707 < 73,737 < 307,988$

5. a. 983,177   b. 555,330

6. a. $3.05 ≈ $3.00   b. $8.32 ≈ $8.00   c. $25.97 ≈ $26.00

7. a. Approximately $130 + $75 + $90 + $140 + $70 = $505.
   b. They earned about $70 less on their worst day than on their best day. (Worst day: about $70, best day about $140.)

8. a. He read $12 × 96 = 1,152$ pages.
   b. Half of the magazines is 6 magazines. Jesse read six magazines in $6 × 2 \ 1/2$ hours = 15 hours.

Cumulative Review: Chapters 1 - 4

1. a. 975 b. 20,990 c. 1,968 d. 4,088

2. $256 + x = 609$; $x = 353$.

3. a. 27 b. 57 c. 22 d. 57 e. 21 f. 1

4. a. $555 \approx 600$ b. $8,889 \approx 8,900$ c. $351,931 \approx 351,900$
   d. $64 \approx 100$ e. $244,295 \approx 244,300$ f. $38,009 \approx 38,000$

5. a. 305,200 b. 40,033

6.

a. $723,050 > 699,099$	b. $322,320 < 322,322$
c. $692,159 < 692,196$	d. $140,000 > 14,100$
e. $113,999 < 115,399$	f. $836,496 > 88,482$

7. a. 1,100; 190; 120,000 b. 180; 800,000; 8,800 c. 92,000; 64,000; 8,800

8.

a. $6 \times 30¢$ $= 180¢ = \$1.80$	b. $5 \times 84¢$ $= 400¢ + 20¢ = \$4.20$
c. $6 \times \$1.70$ $= \$6.00 + \$4.20 = \$10.20$	d. $3 \times \$2.80$ $= \$6.00 + \$2.40 = \$8.40$

9. a. about $20 \times 40 = 800$ plants
   b. The cost was $7 \times \$8.20 = \$56 + \$1.40 = \$57.40$. His change was $\$100 - \$57.40 = \$42.60$.

Cumulative Review: Chapters 1 - 5

1. a. 1,000 meters are not accessible by boat. (Each little "block" in the diagram is 200 m.)

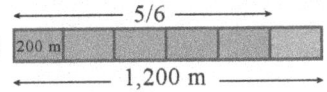

   b. There are 22 girls. There are 33 boys and girls.

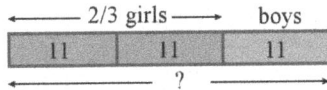

2. a. The baker spent about $90 more for flour in May than in March. In May, he spent about $550 and in March, about $460.
   b. Estimates may vary. He spent about $460 + $620 + $550 = $1,630.

3. a. One triangle weighs 3 units. Solution: First take off two triangles from both sides. That leaves: 11 = 8 + triangle. So, one triangle has to equal 3.

   b. One square weighs 2 units. Solution: Take off one square from both sides. That leaves: 3 squares + 9 = 15. Now, take off "9" from both sides. That leaves 3 squares = 6. So, one square has to weigh 2.

4. a. 1 kg 300 g = 1,300 g   b. 3 lb = 48 oz   c. 7,500 g = 7 kg 500 g
      4 kg 20 g = 4,020 g   7 lb = 112 oz   4 lb 8 oz = 72 oz

5.

Minutes	1	5	6	7	10
Seconds	60	300	360	420	600

Days	1	3	6	10
Hours	24	72	144	240

6. First do 5 + 39 , which equals 44 . Then, divide that answer by 4 .

   This leaves 11 . Then, do 2 × 2 = 4 .

   Lastly, subtract that from 11 . The answer is 7 .

7. a. 4 h 54 m   b. 7 h 24 m   c. 8 h 24 m

8.

a.	b.	c.	d.
2 pt = 4 C	1 qt = 4 C	6 L = 6,000 ml	2 L 560 ml = 2,560 ml
2 C = 16 oz	2 gal = 8 qt	1/4 L = 250 ml	1,300 ml = 1 L 300 ml

9. a. January, February, March, April, September, October, November, and December (the temperature is below freezing, or below 32°).
   b. November, December, and January
   c. 37°

Cumulative Review: Chapters 1 - 6

1. a. $18 + x = 33$; $x = 15.   The unknown x is how much Dana earned.
   b. $100 - $86 = x$; $x = 14. The unknown x is Dad's change.
   c. $120 - 39 = x$; $x = 81$.   The unknown x is the number of eggs that broke.
   d. $13 + 43 = x$; $x = 56$.   The unknown x is the number of dogs the shelter had initially.

2. a. 590. Check: $590 \times 3 = 1{,}770$ b. 878. Check: $878 \times 9 = 7{,}902$.

3. a. 5 R3; 2 R9   b. 5 R3; 3 R8   c. 9 R2; 11 R1

4.

a. 7×78	b. 13×67	c. 311×8
$\approx 7 \times 80 = 560$	$\approx 13 \times 70 = 910$ OR $\approx 10 \times 70 = 700$	$\approx 300 \times 8 = 2{,}400$ OR $\approx 310 \times 8 = 2{,}480$

5. a. $60{,}000 + 70 = 60{,}070$ b. $123{,}000 + 4{,}000 + 4 = 127{,}004$

   c. $3 + 90{,}000 + 40 = 90{,}043$ d. $7 + 20 + 632{,}000 = 632{,}027$

6.

a. 7 m = 700 cm 69 mm = 6 cm 9 mm	b. 2 m 6 cm = 206 cm 6 km = 6,000 m	c. 4 km 100 m = 4,100 m 169 cm = 1 m 69 cm

7.

a. 3 lb 8 oz = 56 oz 4 kg 11 g = 4,011 g	b. 32 oz = 2 lb 4,900 g = 4 kg 900 g	c. 7 lb 2 oz = 114 oz 36 kg 140 g = 36,140 g

8. a. $205   b. 2 km 600 m   c. She spent $7.12. Her change was $2.88

1.

a. $22,934 + 5,312 + 424,787$ Estimation: $23,000 + 5,000 + 420,000 = 448,000$ Calculation: $453,033$	b. $519,313 - 47,616$ Estimation: $520,000 - 50,000 = 470,000$ Calculation: $471,697$

2. $45 \times 22 + 27 = 1,017$ students

3. Right angles are exactly 90°.
   Right triangles have exactly one right angle.

   Obtuse angles are more than 90°, but less than 180°.
   Obtuse triangles have exactly one obtuse angle.

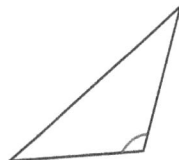

   Acute angles are less than 90°.
   Acute triangles have three acute angles.

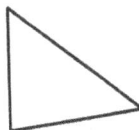

4. Answers will vary. To find possible side lengths, remember that the two side lengths add up to 14 in.

One side	Other side	Perimeter	Area
2 in.	12 in.	28 in.	24 sq. in.
3 in.	11 in.	28 in.	33 sq. in.
4 in.	10 in.	28 in.	40 sq. in.
5 in.	9 in.	28 in.	45 sq. in.

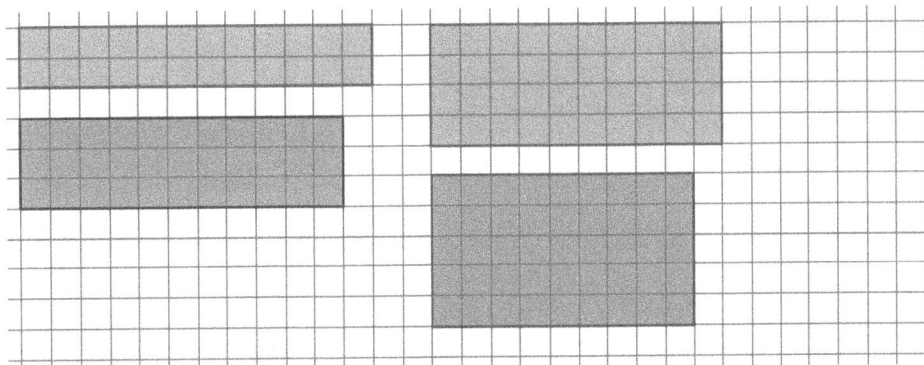

5.

a. 1:40 p.m.
   13 : 40
	b. 9:20 p.m.
	   21 : 20
		c. 2:15 p.m.
		   14 : 15
			d. 10:04 a.m.
			   10 : 04

6.

number	divisible			number	divisible			number	divisible		
	by 2	by 5	by 10		by 2	by 5	by 10		by 2	by 5	by 10
478	x			1,492	x			904	x		
540	x	x	x	3,093				905		x	
255		x		94	x			906	x		

7. No, because the division leaves a remainder: 549 ÷ 7 = 78 R3

8. a. 1, 2, 3, 4, 6, 8, 12, 24   b. 1, 2, 3, 6, 11, 22, 33, 66
   c. 1, 2, 3, 4, 6, 8, 12, 16, 24, 32, 48, 96   d. 1, 3, 5, 15, 25, 75

9.

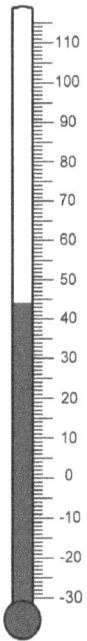

a. 44°F
a chilly fall day

b. 93°F
a hot day

c. 77°F
inside a house

d. 104°F
fever

e. 13°F
a winter day

Cumulative Review: Chapters 1 - 8

1.

a. 4 × 36	b. 5 × 65	c. 8 × 426
$120 + 24 = 144$	$300 + 25 = 325$	$3,200 + 160 + 48 = 3,408$

2. 98,889

3. Answers will vary since estimations can be done in various ways.

a. 8 × 69	b. 11 × 55	c. 25 × 17
$\approx 8 \times 70 = 560$	$\approx 10 \times 60 = 600$ OR $11 \times 60 = 660$ OR $10 \times 55 = 550$	$\approx 25 \times 20 = 500$ OR $24 \times 20 = 480$ (better, rounding one down, one up)

4. a. $x \div 8 = 7$; $x = 56$ b. $24 \div x = 8$; $x = 3$

5. yes - no - no - yes - no - yes - no

6.

a. 5 ft = 60 in.	b. 3 ft 4 in. = 40 in.	c. 4 yd = 12 ft
12 ft = 144 in.	6 ft 6 in. = 78 in.	9 yd = 27 ft

7. a. 5 kg 500 g   b. 3 kg 400 g   c. 9,900 g

8. a. Their total weight was 1 lb 1 oz. 3 oz + 3 oz + 5 oz + 2 oz + 4 oz = 17 oz = 1 lb 1 oz.

   b. 44 boxes. The division is 175 ÷ 4 = 43 R3. Keep in mind, she needs to pack the "leftover" 3 kg also into one box.

   c. She has $84 left. $98 ÷ 7 = 14; $98 − $14 = $84.

   d. Either 2, 3, 4, 6, 9, 12, or 18 children in a row. Probably 4, 6, and 9 children in a row are most practical.

   e. There are 34 foals, and 102 horses in total.

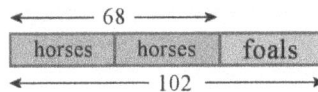

9. Check students' work. a.   b.

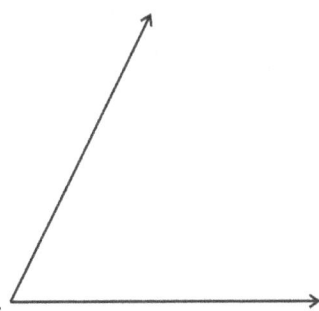

10. $o \parallel AB$, $o \parallel m$, $AB \parallel m$. There are no perpendicular lines, rays, or line segments.

11. Three-fourths of it are left.

12. a. 3   b. 1 1/12   c. 13/100   d. 1 1/4   e. 26/100   f. 6 1/10

13. a. 3   b. 1 1/5   c. 3 3/4   d. 7

14.

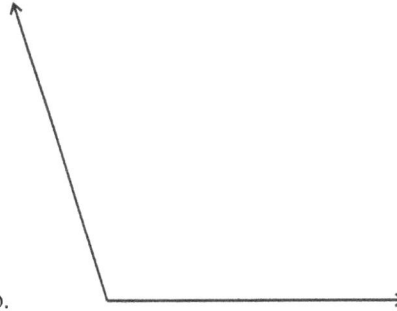

a. $\dfrac{4}{5} = \dfrac{8}{10}$	b. $\dfrac{1}{3} = \dfrac{3}{9}$	c. $\dfrac{2}{3} = \dfrac{8}{12}$

15. a. <   b. <   c. <   d. >